Lycium Barbarum and Human Health

Raymond Chuen-Chung Chang • Kwok-Fai So
Editors

Lycium Barbarum and Human Health

 Springer

Editors
Raymond Chuen-Chung Chang
Laboratory of Neurodegenerative Diseases
Department of Anatomy
LKS Faculty of Medicine the University
of Hong Kong
Hong Kong
China

Kwok-Fai So
Department of Anatomy
LKS Faculty of Medicine the University
of Hong Kong
Hong Kong
China

ISBN 978-94-017-9657-6   ISBN 978-94-017-9658-3 (eBook)
DOI 10.1007/978-94-017-9658-3

Library of Congress Control Number: 2014958968

Springer Dordrecht Heidelberg New York London

Springer is part of Springer Science+Business Media (www.springer.com)

Preface

The fruit of *Lycium barbarum* (also called Wolfberry) is known to be anti-aging and nurtures the eyes or vision. It is an upper class Chinese medicine, meaning that it can be used as medicine for therapy as well as an ingredient in Chinese cuisine. Collective efforts from different research teams have proven that the fruits of *Lycium barbarum* have rich sources that protect our whole body, from the skin to the liver, brain and eyes. Therefore, regular consumption of *Lycium barbarum* can help us to keep the balance of Yin/Yang in our body to fight against any possible age-associated diseases.

There is a famous Chinese story related to *Lycium barbarum*. One day a young man was walking in a village. On his way, he found two people arguing with each other in a narrow lane. He went to see what had happened and found a relatively young and strong man with black hair arguing with a weak elder with grey hair. It looked like the elder had been blamed for something. In Chinese culture, we all have great respect for the elderly. This was why this young man did not feel quite right and thought that the man with black hair was not paying respect to his senior. He asked this weak elder with grey hair whether the black-haired man had done him some injustice. The grey-haired man then pointed to the strong black-haired man, saying, 'he is my big brother'. The strong black-haired big brother said that his little brother did not listen to him and take *Lycium barbarum*. This was why his little brother looked old and weak.

From this story, we have insight into the beneficial effects of *Lycium barbarum*. In this book, we have carefully arranged the content from the plant, the chemical components and the effects on different organs/biological systems as well as its potential harmful effects. Authors in every chapter used different scientific methods to prove the effects of *Lycium barbarum*. We are not just showing the benefits of *Lycium barbarum*. Some people may be allergic to *Lycium barbarum*. This book is the first book about *Lycium barbarum* written in English. As more people are searching for health food supplements and there are many so-called 'secrete formulations of herbs and health food supplements', we should look for some reliable health food with solid scientific evidence and be cautious of any possible deleterious effects. We hope that this book gives us a comprehensive understanding of the pros and cons of this anti-aging *Lycium barbarum*.

Raymond Chuen-Chung Chang
Kwok-Fai SO   v

Contents

Contributors

María Benlloch Universidad Católica de Valencia 'San Vicente Mártir', Valencia, Spain

Krzysztof Bojanowski Sunny BioDiscovery, Inc., Santa Paula, CA, USA

Jerónimo Carnés R&D Department, Laboratorios LETI S.L., Madrid, Spain

Gloria Castellano Universidad Católica de Valencia 'San Vicente Mártir', Valencia, Spain

Henry HL Chan Laboratory of Experimental Optometry (Neuroscience), School of Optometry, The Hong Kong Polytechnic University, Hong Kong, People's Republic of China

Raymond Chuen-Chung Chang Laboratory of Neurodegenerative Diseases Department of Anatomy, The University of Hong Kong, Hong Kong, China

GHM Institute of CNS Regeneration and Guangdong Key Laboratory of Brain Function and Diseases, Jinan University, Guangzhou, China

Research Centre of Heart, Brain, Hormone and Healthy Aging, LKS Faculty of Medicine, The University of Hong Kong, Hong Kong, People's Republic of China

Hao Chen South China Botanical Garden, Chinese Academy of Sciences, Guangzhou, Guangdong Province, China

Huilin Chen Ningbo College of Health Sciences, Ningbo, People's Republic of China

Patrick HW Chu Laboratory of Experimental Optometry (Neuroscience), School of Optometry, The Hong Kong Polytechnic University, Hong Kong, People's Republic of China

Yong Ding Department of Ophthalmology, The First Affiliated Hospital of Jinan University, Guangzhou, Guangdong, China

Jingzhou Dong Wuhan Botanical Garden, Chinese Academy of Sciences, Wuhan, Hubei Province, China

Angel Ferrer   Allergy Unit, Hospital General Universitario de Elche, Alicante, Spain

Miguel Flores-Bellver   Universidad Católica de Valencia 'San Vicente Mártir', Valencia, Spain

Emilio González-García   Universidad Católica de Valencia 'San Vicente Mártir', Valencia, Spain

Junming Guo   Department of Biochemistry and Molecular Biology, and Zhejiang Provincial Key Laboratory of Pathophysiology, Ningbo University School of Medicine, Ningbo, People's Republic of China

Clara Hiu-Ling Hung   Laboratory of Neurodegenerative Diseases, Department of Anatomy, LKS Faculty of Medicine, The University of Hong Kong, Hong Kong SAR, People's Republic of China

Institute of Chinese Medicinal Science, University of Macau, Macau SAR, People's Republic of China

Yuen-Shan Ho   School of Nursing, Faculty of Health and Social Sciences, The Hong Kong Polytechnic University, Hong Kong SAR, People's Republic of China

Ruo-Jing Huang   Department of Ophthalmology, The First Affiliated Hospital of Jinan University, Guangzhou, Guangdong, China

Julio Huertas   Allergy Section, Complejo Hospalario Universitario de Cartagena, Murcia, Spain

Carlos H. de Larramendi   Allergy Section, Hospital Marina Baixa, Villajoyosa and Centro de Especialidades Foietes, Benidorm, Alicante, Spain

Benson Wui-Man Lau   Department of Rehabilitation Sciences, The Hong Kong Polytechnic University, Hong Kong, People's Republic of China

LKS Faculty of Medicine, Department of Anatomy, The University of Hong Kong, People's Republic of China, Hong Kong

Mason Chin-Pang Leung   Department of Rehabilitation Sciences, The Hong Kong Polytechnic University, Hong Kong, People's Republic of China

Hong-Ying Li   GHM Institute of CNS Regeneration and Guangdong Key Laboratory of Brain Function and Diseases, Jinan University, Guangzhou, People's Republic of China

Department of Anatomy, Jinan University School of Medicine, Guangzhou, People's Republic of China

Department of Ophthamology, The University of Hong Kong, Hong Kong, People's Republic of China

The State Key Laboratory of Brain and Cognitive Science and the Research Centre of Heart, Brain, Hormone and Healthy Aging, The University of Hong Kong, Hong Kong, People's Republic of China

Peifei Li   Department of Biochemistry and Molecular Biology, and Zhejiang Provincial Key Laboratory of Pathophysiology, Ningbo University School of Medicine, Ningbo, People's Republic of China

Xiao-ang Li   State Key Laboratory of Quality Research in Chinese Medicine, Macau University of Science and Technology, Macau SAR, People's Republic of China

Yongliang Liu   Wuhan Botanical Garden, Chinese Academy of Sciences, Wuhan, Hubei Province, China

Amy CY Lo   Department of Ophthalmology, Li Ka Shing Faculty of Medicine, The University of Hong Kong, Hong Kong, People's Republic of China

María Angeles López-Matas   R&D Department, Laboratorios LETI S.L., Madrid, Spain

Xue-Song Mi   Department of Ophthalmology, The First Affiliated Hospital of Jinan University, Guangzhou, Guangdong, China

Department of Anatomy, The University of Hong Kong, Hong Kong, China

María Muriach   Unidad predepartamental Medicina, Universitat Jaume I, Castellón de la Plana, Spain

Kai-Ting Po   Department of Rehabilitation Sciences, The Hong Kong Polytechnic University, Hong Kong, People's Republic of China

Francisco J. Romero   Universidad Católica de Valencia 'San Vicente Mártir', Valencia, Spain

Department of Physiology, University CEU–Cardenal Herrera, Valencia, Spain

Department of Physiology, School of Medicine, Universidad Católica de Valencia 'San Vicente Mártir', Valencia, Spain

Francisco Javier Sancho-Pelluz   Universidad Católica de Valencia 'San Vicente Mártir', Valencia, Spain

Kwok-Fai So   Department of Anatomy, The University of Hong Kong, Hong Kong, China

GHM Institute of CNS Regeneration and Guangdong Key Laboratory of Brain Function and Diseases, Jinan University, Guangzhou, China

Department of Ophthalmology, The University of Hong Kong, Hong Kong, China

The State Key Laboratory of Brain and Cognitive Science, The University of Hong Kong, Hong Kong, People's Republic of China

LKS Faculty of Medicine, Research Centre of Heart, Brain, Hormone and Healthy Aging, The University of Hong Kong, Hong Kong, People's Republic of China

George L. Tipoe   Department of Anatomy, The University of Hong Kong, Hong Kong SAR, People's Republic of China

Ying Wang   South China Botanical Garden, Chinese Academy of Sciences, Guangzhou, Guangdong Province, China

Wuhan Botanical Garden, Chinese Academy of Sciences, Wuhan, Hubei Province, China

Min Wu   South China Botanical Garden, Chinese Academy of Sciences, Guangzhou, Guangdong Province, China

Bingxiu Xiao   Department of Biochemistry and Molecular Biology, and Zhejiang Provincial Key Laboratory of Pathophysiology, Ningbo University School of Medicine, Ningbo, People's Republic of China

Jia Xiao   Department of Immunobiology, Institute of Tissue Transplantation and Immunology, Jinan University, Guangzhou, People's Republic of China

Di Yang   Department of Ophthalmology, Li Ka Shing Faculty of Medicine, The University of Hong Kong, Hong Kong, People's Republic of China

Shaohua Zeng   South China Botanical Garden, Chinese Academy of Sciences, Guangzhou, Guangdong Province, China

Xiaorui Zhang   State Key Laboratory of Toxicology and Medical Countermeasures, Beijing Institute of Pharmacology and Toxicology, Beijing, People's Republic of China

Yongxiang Zhang   State Key Laboratory of Toxicology and Medical Countermeasures, Beijing Institute of Pharmacology and Toxicology, Beijing, People's Republic of China

Hui Zhao   Sunny BioDiscovery, Inc., Santa Paula, CA, USA

Wenxia Zhou   State Key Laboratory of Toxicology and Medical Countermeasures, Beijing Institute of Pharmacology and Toxicology, Beijing, People's Republic of China

Chapter 1
Chemical and Genetic Diversity of Wolfberry

Ying Wang, Hao Chen, Min Wu, Shaohua Zeng, Yongliang Liu
and Jingzhou Dong

Abstract *Lycium* (Boxthorn) is a genus of the nightshade family (Solanaceae), containing about 80 species of plants native throughout the temperate and subtropical zones of the world. Wolfberry nowadays in China refers to the products prepared from *Lycium chinense*, *Lycium barbarum*, and *Lycium ruthenicum*. Wolfberry has been consumed as food and medicine for more than 4000 years, and the cultivation of *L. barbarum* has been recorded for more than 600 years in the Northwestern part of China, especially Ningxia province which is also the authentic region of Chinese medicine, Lycii Fructus. This review will cover the history, cultivation, genetic diversity, and phytochemical diversity of these three species. High level of genetic diversity has been discovered in wild resources. Phytochemical diversity includes polysaccharides, carotenoids, flavonoids, alkaloids, amides, peptides, anthraquinones, coumarins, lignanoids, terpenoids, sterols, steroids, organic acids, anthocyanins, essential oils, glycolipids, and others from leaves, fruits, and root bark of *L. chinense*, *L. barbarum*, and *L. ruthenicum*.

Keywords Genetic diversity · Phytochemical diversity · History · *L. chinense* · *L. barbarum* · *L. ruthenicum*,

1.1 Introduction

Lycium (Boxthorn) is a genus of the nightshade family (Solanaceae), containing about 80 species of plants native throughout the temperate and subtropical zones of the world (Levin et al. 2011). They are mostly found in dry, semisaline

Y. Wang (✉) · H. Chen · M. Wu · S. Zeng
South China Botanical Garden, Chinese Academy of Sciences, Guangzhou,
Guangdong Province, China
e-mail: yingwang@wbgcas.cn

Y. Wang · Y. Liu · J. Dong
Wuhan Botanical Garden, Chinese Academy of Sciences, Wuhan, Hubei Province, China

© Springer Science+Business Media Dordrecht 2015   1
R. C-C. Chang, K-F. So (eds.), *Lycium Barbarum and Human Health*,
DOI 10.1007/978-94-017-9658-3_1

environments. Other common names include Wolfberry, Goji, Desert-thorn, Christmas berry, Matrimony vine, and Duke of Argyll's tea tree, as well as *Gouqi* in Chinese. Chinese Pharmacopoeia (2010) recorded Lycii Fructus (*Gouqizi*, dry fruit of *Lycium barbarum*), and Lycii Cortex (*Digupi*, dry root bark of *Lycium chinense* and *L. barbarum*). Young leaves of *L. chinense* and *L. barbarum* are consumed as functional vegetable and functional tea. A wide range of wolfberry products have been developed, including tea, wine, cosmetic products, milk, coffee, juice, seed oil, etc. Dark purple color fruits of *Lycium ruthenicum*, also called *"Hei guo gou qi"* or black fruit wolfberry, have been used as folk medicine, especially as Tibetan and Mongolian medicine. Therefore, wolfberry nowadays in China refers to the products prepared from *L. chinense*, *L. barbarum*, and *L. ruthenicum*. Only *"Ning Xia Gou Qi"* (later on refer as Goji or Goji berry) in Chinese indicates specifically fruits of *L. barbarum*. Wolfberry has been consumed as food and medicine for more than 4000 years, and the cultivation of *L. barbarum* has been recorded for more than 600 years in the Northwestern part of China, especially Ningxia province which is also the authentic region of Chinese medicine, Lycii Fructus. This chapter will cover the history, genetic diversity, and phytochemical diversity of these three species.

1.2   History of Wolfberry

Mr. Shizhen Li, the great pharmaceutical scientist in the Ming Dynasty, described the origin of the name Gouqi (later on refer as Goji) in his famous book *Compendium of Materia Medica* (*Ben Cao Gang Mu*). He explained that "Gouqi" was a combination of "Gou" and "Qi" as the thorns of the Gouqi tree were like the ones of the "Gou" tree while the stems of the Gouqi tree were like the branches of the "Qi" tree. Goji has been in the Chinese culture for a long time. Its earliest record was found in the oracle bone script (Jiaguwen) of the Shang Dynasty, which indicates that Goji may have been recognized, cultivated, and utilized as early as in the Xia Dynasty which occurs about 4000 years ago.

1.2.1   The Medicinal and Culinary Culture of Goji

Goji berries have long served as a good herbal tonic. There are many instances in Chinese history in which people become healthier and longer lived due to the tonic effect of Goji. Goji berry is one of the constituents of the three prescribed medicines with secret recipes in the period of the First Emperor of the Qin Dynasty (Qin Shihuang). Fang Xuanling, a famous prime minister in the Tang Dynasty once became exhausted in his body and mind due to overwork. However, he recovered well by keeping having the wolfberry tremella soup. In addition, Goji berries serve as one of the important constituents of two tonics fed by the Empress Dowager Cixi of the Qing Dynasty.

The medicinal effect of Goji berries has long been recognized and appreciated by numerous medical scientists in the history of China. *Shennong's Root* and *Herbal Classic* (*Shennong bencao jing*), a classic work on medicinal plants and their uses, is the oldest pharmaceutical work existing in China. This work was written and published in the periods from the Warring States to the Han Dynasty and recorded and summarized the medical and pharmaceutical knowledge in ancient China. In the 365 medicines recorded in this work, Goji berries fell into the top-class medicines which can contribute much to longevity. Based on the recorded research on Goji in the history, Li shizhen summarized the healthy and tonic effect of Goji in *Compendium of Materia Medica*. For instance, he found that many organs other than the Goji berries also possessed healthy and tonic effect. Therefore, different organs can be collected for medicinal use at different seasons. Specifically, the leaf, flower, fruit, and root of Goji can be collected in the spring, summer, autumn, and winter, respectively.

1.2.2 The Literature About Goji

Goji has not only been used in cooking and health care, but also serving as the topic of many literatures like poems written in the palace banquets or the life in ancient China. For example, *Classic of Poetry* is the oldest existing collection of Chinese poetry comprising works in the periods from the Western Zhou Dynasty to the Spring and Autumn period dating from the eleventh to seventh centuries BC. There are seven poems involving Goji in this collection, six of which are in the poetry *Xiaoya*. Besides, many famous poets in ancient China, like Du Pu, Bai Juyi, and Liu Yuxi in the Tang Dynasty, Su Shi, Lu You, and Huang Tingjian in the Northern Song Dynasty, once wrote poems to appreciate Goji.

1.2.3 Zhongning: The Hometown of Goji in China

Goji has long been cultivated in the Ningxia Hui Autonomous Region in China. In the poem *Beishan* collected in *Xiaoya*, the author described experience to collect Goji berries in a place called "Beishan". It has been verified that this place corresponds to several counties in the Zhongwei City currently, including the Haiyuan County and the Zhongning County, which indicates a long history for Goji cultivation in these places.

The unique geographic environment and regional climate in the Zhongning County offer the most exceptional natural environment for cultivating Goji. Specifically, there is sufficient sunlight, high effective accumulated temperatures and large temperature difference between day and night; the soil in the alluvial plain contains a tremendous amount of minerals and plenty of humus and is mature enough; the Goji plants can be easily irrigated with the water of unique quality. Grown in such excellent natural environment, the Goji berries produced in Ningxia, especially in Zhongning, are renowned worldwide for their exceptional quality.

Actually, during the periods from the Ming Dynasty to Republic of China, there is evidence that the Goji berries produced in Zhongning once dominated the market in China. For instance, it was recorded by a county head named Huang Enxi in the *Zhongwei Annals* that there were many farmers growing Goqi in the area of Zhongning and the medicinal Goji berries in various provinces of China were all produced in Zhongning.

In 2013, the cultivation area of Goji in Zhongning has increased to about 13,000 ha, with the output value for dried berries reaching 1 billion yuan (RMB). Besides, the Goji berries produced in Zhongning has won plenty of honors. For instance, in 1961, the Zhongning County was nominated as the only national base to produce Goji berry by The State Council of the People's Republic of China; in 1995, it was named "the hometown of Goji berry in China" by The State Council; in 1999, the Goji berry from Zhongning won the gold medal in the International Horticultural Exposition held in Kunming, China.

1.3   Identification of *L. chinense, L. barbarum, L. ruthenicum*

L. barbarum L. (*Ning xia gou qi* in Chinese), *L. chinense* Miller (*Gouqi*), and *L. ruthenicum* Murray (*Hei guo gou qi*) are currently the three major *Lycium* species consumed in China as functional vegetable and fruit, Chinese medicine, folk medicine, and food supplements. The morphology of plants of these species shares certain common characters (Zhang et al. 1994). For instances, plants are all shrubs of 20–200 cm in height; branches are thorny; leaves are solitary or fasciculate while leaf size is about 0.5–5 cm × 0.5–7 mm; inflorescences are solitary or clustered; calyx is campanulate, 3–5 mm in length and 2–5-lobed; corolla is purple and funnel form. However, these three species display certain morphological differences, and the natural habitats are overlapped but different (Table 1.1). Leaf, flower, and fruit have major morphological characters that can distinguish these three species (Table 1.1, Fig. 1.1). Although both have red color berry fruits, *L. barbarum* has longer and thicker leaves than that of *L. chinense*. *L. ruthenicum* has subsessile and succulent leaves and purple-black berry (Fig. 1.1) (Zhang et al. 1994).

Table 1.1. Morphological differences and geographic distribution of *L. chinense*, *L. barbarum*, *L. ruthenicum*

Species	Morphological identity	Geographic distribution
L. chinense M	Leaf blade ovate, rhombic, lanceolate, or linear-lanceolate. Pedicel 1–2 cm. Calyx 3–5-divided to halfway, lobes densely ciliate. Corolla tube shorter than or sub-equaling lobes, lobes pubescent at margin. Stamens filaments villous slightly above base. Berry red, ovoid or oblong. Seeds numerous, yellow	In China: Anhui, Fujian, Gansu, Guangdong, Guangxi, Guizhou, Hainan, Hebei, Heilongjiang, Henan, Hubei, Hunan, Jiangsu, Jiangxi, Jilin, Liaoning, Nei Mongol, Ningxia, Qinghai, Shaanxi, Shanxi, Sichuan, Yunnan, Zhejiang. Taiwan, Japan, Korea, Mongolia, Nepal, Pakistan, Thailand, SW Asia, Europe
L. barbarum L	Leaves lanceolate or long elliptic. Pedicel 1–2 cm. Calyx usually 2-lobed, lobes 2- or 3- toothed at apex. Corolla tube 8–10 mm, obviously longer than limb and lobes; lobes 5–6 mm, spreading, margin glabrescent. Berry red or orange-yellow, oblong or ovoid. Seeds usually 4–20, brown-yellow	In China: Ningxia, Gansu, Qing-hai, Xinjiang
L. ruthenicum M	Shrubs copiously armed. Leaves subsessile; leaf succulent, linear or subcylindric, rarely linear-oblanceolate. Pedicel 5–10 mm. Calyx irregularly 2–4-lobed, lobes sparsely ciliate. Corolla lobes oblong ovate, not ciliate. Stamens filaments sparsely villous above base. Fruiting calyx slightly inflated. Berry purple-black, globose, sometimes emarginate. Seeds brown	In China: Gansu, Nei Mongol, Ningxia, Qinghai, N Shaanxi, Xinjiang, Xizang. Afghanistan, Kazakhstan, Kyrgyzstan, Mongolia, Pakistan, Russia, Tajiki-stan, Turkmenistan, Uzbekistan, SW Asia, Europe

Fig.1.1 Ripen fruits of *L. ruthenicum* (**a**), *L. barbarum* (**b**), and *L. chinense* (**c**)

1.4   Chemical Diversity in Leaf and Fruits of Wolfberry

Polysaccharides represent quantitatively the most important group of substances in the fruit of *L. barbarum*. which are estimated to comprise 3–8 % of the dried fruits (Amagase and Farnswoeth 2011). More than 30 polysaccharides have been isolated from the fruit of *L. barbarum*, *L. chinense*, and *L. ruthenium* (Table 1.2). The molecular weight of polysaccharides varies greatly in different species, which might lead to the different pharmacological function.

Table 1.2 Polysaccharides in fruits of *L. chinense*, *L. barbarum*, *L. ruthenicum*

Glycocon-jugate	MW	Carbo-hydrate content	Monosaccharides (molar ratio or %)	Reference
L. barbarum				
LbGp1	88,000		Ara, Gla, Glc (2.5:1.0:1.0)	(Yao et al. 2011)
LbGp2	68,200	90.7	Ara, Gal (4:5)	(Peng and Tian 2001)
LbGp3	92,500	93.6	Ara, Gal (1:1)	(Huang et al. 1998, 1999)
LbGp4	214,800	85.6	Ara, Gal, Rha, Glc (1.5:2.5:0.43:0.23)	(Huang et al. 1998; Peng et al. 2001a)
LbGp5	23,700	8.6	Rha, Ara, Xyl, Gal, Man, Glc (0.33:0.52:0.42:0.94:0.85:1)	(Huang et al. 1998)
LbGp5B	23,700		Rha, Ara, Glc, Gal, (0.1:1:1.2:0.3), Galu (0.9)	(Peng et al. 2001b)
LBP3p	157,000	92.4	Gal, Glc, Rha, Ara, Man, Xyl (1:2.12:1.25:1.10:1.95:1.76)	(Gan et al. 2004)
LBPC2	12,000	92.8	Xyl, Rha, Man (8.8:2.3:1)	(Zhao et al. 1996, 1997)
LBPC4	10,000	95	Glc	(Zhao et al. 1996, 1997)
LBPA1	18,000		Heteroglycan	(Zhao et al. 1997)
LBPA3	66,000		Ara, Gal (1.2:1)	(Zhao et al. 1997)
LBP1a-1	11,500		Glc	(Duan et al. 2001)
LBP1a-2	9400		Glc	(Duan et al. 2001)
LBP3a-1	10,300		GalA	(Duan et al. 2001)
LBP3a-2	8200		GalA	(Duan et al. 2001)
LBPF1	150,000	48.2		(Chen et al. 2008)
LBPF2	150,000	30.5		(Chen et al. 2008)
LBPF3	150,000	34.5		(Chen et al. 2008)
LBPF4	150,000	20.3		(Chen et al. 2008)
LBPF5	150,000	23.5		(Chen et al. 2008)
LBPB1	18,000		Ara, Glc (1:3.1)	(Zhao et al. 1996, 1997)
PLBP	121,000			(Liang et al. 2011)
LBP-IV	418,000		Rha, Ara, xyl, glc, Gal (1.61:3.82:03.44:7.54:1.00)	(Liu et al. 2012)
L. chinense				
Cp-1-A	10,000	87.8	Ara, Xyl (1:1)	(Qin et al. 2000)
Cp-1-B	11,000	89.4	Ara	(Qin et al. 2000)
Cp-1-C	42,000	92.4	Ara, Gal (3:1)	(Qin et al. 2000)
Cp-1-D	23,000	90.7	Ara, Gal (1:1)	(Qin et al. 2000)
Cp-2-A	89,000	88.3	Ara (50.6), Gal (22.8), Man (8.4), Rha (5.9), Glc (5.6)	(Qin et al. 2001)
Cp-2-B	89,000	88.3	Ara (45.5), Gal (47.4)	(Qin et al. 2001)
Hp-2-A	8000	87.9	Ara (70.6), Gal (13.5)	(Qin et al. 2001)

Table 1.2 (continued)

Glycocon-jugate	MW	Carbo-hydrate content	Monosaccharides (molar ratio or %)	Reference
Hp-2-B	11,000	89.9	Ara (84.2), Gal (10.7)	(Qin et al. 2001)
Hp-2-C	120,000	90.7	Ara (49.5), Gal (40.8), Fuc (5.9)	(Qin et al. 2001)
Hp-0-A	23,000		Ara	(Potterat 2010)
L. ruthenicum				
LRGP1	56,200		Rha, Ara, xyl, Man, glc, Gal (0.65:10.71:0.33:0.67:1:10.41)	(Peng et al. 2012a)
LRP4-A	105,000		Rha, Ara, glc, Gal (1:7.6:0.5:8.6)	(Lv et al. 2013)
LRGP3	75,600		Rha, Ara, Gal (1.0:14.9:10.4)	(Peng et al. 2012b)

As to the color regents in wolfberry, the reddish-orange color of *L. barbarum* and *L. chinense* is derived from carotenoids and their esters, which are the second major group of metabolites. Twelve carotenoids and their esters were identified in the genus *Lycium* (Table 1.2.). The highest content of carotenoids in ripen red berry is zeaxanthin dipalmitate which counts for 75 % of total carotenoids (508.90 μg g^{-1} fresh weight (FW) in ripen fresh fruit) (Liu et al. 2014). Although there is very low level of total carotenoid (34.46 μg g^{-1} FW), with 18.01μg g^{-1} FW of β-carotene, in green fruits of *L. ruthenicum*, the content of total carotenoid in ripen black berry is undetectable (Liu et al. 2014). The zeaxanthin and β-Cryptoxanthin are undetectable both in green and ripen fruits of *L. ruthenicum* (Liu et al. 2014). As to the black color in ripen fruits of *L. ruthenicum*, ten anthocyanins were identified using HPLC-DAD-MS/MS (Zheng et al. 2011), with the highest content of pentunidin-3-O-rutinoside (trans-p-coumaroyl)-5-O-glucoside which counts 95 % of total flavonoids (Zeng et al. 2014). Consistent with this, anthocyanin content in *L. ruthenicum* increased steadily and reached maximum levels (10.37 OD$_{534}$/g) at the ripening stage, while anthocyanin was undetectable at all stages in *L. barbarum* fruits (Zeng et al. 2014).

Other phytochemicals include flavonoids, alkaloids, amides, peptides, anthraquinones, coumarins, lignanoids, terpenoids, steroids, and their derivatives, organic acids, and glycolipids are summarized in Table 1.3. Kim et al. (1997c) identified 45 volatile flavor components in *L. chinense* leaves including four acids, 15 alcohols, seven aldehydes, two esters, three furans, nine hydrocarbons, and three others. Sannai et al. (1983) identified 36 neutral volatile compounds in *L. chinense* fruits. Fifty-four volatile components including twelve alcohols, twelve esters, seven aldehydes, six acids, five hydrocarbons, eight ketones, one furan, and three pyrazines were detected in the fruit of *L. chinense* (Yao et al. 2011). Twenty-one compounds from the essential oil of *L. barbarum* fruits and 18 compounds from the essential oil of *L. ruthenicum* fruits were identified by GC/MS (Altintas et al. 2006). 1β-Amino-3β, 4β, 5α-trihydroxycycloheptane, digupigan A, and a tryptophane glycoside, were only isolated from the root barks of *L. chinense* (Asano et al. 1997; Yahara et al. 1989; Wei and Liang 2003). The only one lignin, (+)-Lyoniresinol 3α-*O*-β-D-

Table 1.3 Chemical constituents of *L. barbarum, L. chinense,* and *L. ruthenicum*

Compound name	*L. barbarum*	*L. chinense*	*L. ruthenicum*
Carotenoids and their esters			
β-Carotene	Fruit (Yao et al. 2011)	Fruit/leaf (Yao et al. 2011)	
Zeaxanthin	Fruit (Yao et al. 2011)	Fruit (Yao et al. 2011)	
β-Cryptoxanthin	Fruit (Yao et al. 2011)	Fruit (Yao et al. 2011)	
Zeaxanthinmonopalmitate	Fruit (Yao et al. 2011)	Fruit (Yao et al. 2011)	
Zeaxanthindipalmitate	Fruit (Yao et al. 2011)	Fruit (Kim et al. 1997b)	
Zeaxanthinmonomyristate	Fruit (Yao et al. 2011)		
Zeaxanthinmyristate/palmitate	Fruit (Yao et al. 2011)		
β-Cryptoxanthinpalmitate	Fruit (Yao et al. 2011)		
Violaxanthindipalmitate	Fruit (Yao et al. 2011)		
Mutatoxanthindipalmitate	Fruit (Yao et al. 2011)		
Antheraxanthindipalmitate	Fruit (Yao et al. 2011)		
Lutein		Fruit/ leaf (Yao et al. 2011)	
Flavonoids			
Quercetin	Fruit/ leaf/flower (Yao et al. 2011)	Fruit/leaf (Miean and Mohamed 2001)	
Kaempferol	Fruit/ leaf/flower (Yao et al. 2011)		
Myricetin	Fruit (Le et al. 2007)		
Rutin	Fruit/leaf (Yao et al. 2011)	Fruit/leaf/root (Yao et al. 2011)	
Isorhamnetin 3-*O*-rutinoside	Fruit (Inbaraj et al. 2010)		
Kaempferol-3-*O*-rutinoside	Fruit (Inbaraj et al. 2010)		
Hesperidin	Fruit (Inbaraj et al. 2010)		
Apigenin		Leaf/root bark (Miean and Mohamed 2001)	
Luteolin		Leaf (Zou 2002)	
Acacetin		Leaf (Zou 2002)	
3, 5, 7, 3′-Tetrahydroxy-6, 4′, 5′-trimethoxyflavone		Leaf (Zou 2002)	
Morin		Fruit (Qian et al. 2004)	
Acatein 7-*O*-rhamno-syl-(1-6)-glucopyranoside		Leaf (Zou 2002)	
Quercetin 3-*O*-sophoroside		Leaf (Yao et al. 2011)	
Quercetin 7-*O*-glucoside 3-*O*-glucosyl-(1-2)-galacto-pyranoside		Leaf (Yao et al. 2011)	

Table 1.3 (continued)

Compound name	*L. barbarum*	*L. chinense*	*L. ruthenicum*
Kaempferol 3-*O*-sophoroside		Leaf (Yao et al. 2011)	
Kaempferol 7-*O*-glucoside 3-*O*-glucosyl-(1-2)-galactoside		Leaf (Yao et al. 2011)	
Linarin		Leaf (Zou 2002; Wei and Liang 2003)	
Alkaloids			
Atropine	Fruit/shoot/root (Harsh 1989; Adams et al. 2006)		
Hyoscyamine	Fruit/ shoot/root (Harsh 1989)		
N^a-[(*E*)-Cinnamoyl] histamine	Leaf (Yao et al. 2011)		
Betaine	Fruit/ leaf/ root bark (Yao et al. 2011)		
Melatonin	Fruit (Yao et al. 2011)		
Calystegine A3		Root bark (Asano et al. 1997)	
Calystegine A5		Root bark (Asano et al. 1997)	
Calystegine A6		Root bark (Asano et al. 1997)	
Calystegine A7		Root bark (Asano et al. 1997)	
Calystegine B1		Root bark (Asano et al. 1997)	
Calystegine B2		Root bark (Asano et al. 1997)	
Calystegine B3		Root bark (Asano et al. 1997)	
Calystegine B4		Root bark (Asano et al. 1997)	
Calystegine B5		Root bark (Asano et al. 1997)	
Calystegine C1		Root bark (Asano et al. 1997)	
Calystegine C2		Root bark (Asano et al. 1997)	
Calystegine N1		Root bark (Asano et al. 1997)	
N-Methylcalystegine B2		Root bark (Asano et al. 1997)	

Table 1.3 (continued)

Compound name	*L. barbarum*	*L. chinense*	*L. ruthenicum*
N-Methylcalystegine C1		Root bark (Asano et al. 1997)	
Fagomine		Root bark (Asano et al. 1997)	
6-Deoxyfagomine		Root bark (Asano et al. 1997)	
4-[2-Formyl-5-(hydroxymethyl)-1*H*-pyrrol-1-yl] butanoic acid		Fruit (Chin et al. 2003)	
4-[2-Formyl-5-(methoxymethyl)-1*H*-pyrrol-1-yl] butanoic acid		Fruit (Chin et al. 2003)	
4-[2-Formyl-5-(methoxymethyl)-1*H*-pyrrol-1-yl] butanoate		Fruit (Chin et al. 2003)	
Alkaloid I		Root bark (Yao et al. 2011)	
Alkaloid I		Root bark (Yao et al. 2011)	
Kukoamine A		Root bark (Funayama et al. 1980)	
Kukoamine B		Root bark (Yao et al. 2011)	
Betaine		Fruit/ leaf/root bark/ root (Yao et al. 2011)	
Betaine hydrochloride		Root bark (Zhou et al. 1996)	
Choline		Root/ root bark (Yao et al. 2011)	
9-Formylharman		Fruit (Han et al. 1985)	
1-(Methoxycarbonyl)-*β*-carboline		Fruit (Han et al. 1985)	
Perlolyrine		Fruit (Han et al. 1985)	
Amides			
Lyciumide A	Fruit (Yao et al. 2011)		
3-(3-Hydroxy-4-methoxyphenyl)-*N*-[2-(4- methoxyphenyl) ethyl]-(2*E*)-Propenamide	Fruit (Personal communication with Dr. Minghua Qiu)		
3-(4-Hydroxy-3-methoxyphenyl)-*N*-[2-(4- hydroxyphenyl) ethyl]-2-Propenamide	Fruit (Personal communication with Dr. Minghua Qiu)		

Table 1.3 (continued)

Compound name	L. barbarum	L. chinense	L. ruthenicum
N-(α,β-Dihydrocaffeoyl) tyramine		Root bark (Han et al. 2002; Lee et al. 2004)	
N-[(E)-Caffeoyl]tyramine		Root bark (Han et al. 2002; Lee et al. 2004)	
N-[(Z)-Caffeoyl]tyramine		Root bark (Han et al. 2002; Lee et al. 2004)	
N-[(E)-Feruloyl] octopamine		Root bark (Lee et al. 2004)	
(2S, 3R, 4E, 8Z)-1-O-(β-D-Glucopyranosyl)-2-(palmitoylamino) octadecasphinga-4,8-diene		Fruit(Kim et al. 1997a, 2000)	
(2S, 3R, 4E, 8Z)-1-O-(β-D-Glucopyranosyl)-2-[(2-hydroxypalmitoyl)amino] sphinga-4,8-diene		Fruit/suspension culture stem (Kim et al. 1997a; Jang et al. 1998)	
Peptides			
Lyciumamide		Stem/root bark (Noguchi et al. 1984)	
Lyciumins A		Root bark (Yahara et al. 1989)	
Lyciumins B		Root bark (Yahara et al. 1989)	
Lyciumins C		Root bark (Yahara et al. 1993)	
Lyciumins D		Root bark (Yahara et al. 1993)	
Anthraquinones			
Emodin		Root bark (Wei and Liang 2002)	
Physcion		Root bark (Wei and Liang 2002)	
1, 3, 6-Trihydroxy-2-methyl-9, 10-anthraqui-none		Root bark (Yao et al. 2011)	
1, 3, 6-Trihydroxy-2-methyl-9, 10-anthraquinone 3-O-(rhamnopyranosyl)-(1-2)-6′-acetylglucopyranoside		Root bark (Yao et al. 2011)	
Coumarins			
Scopoletin	Fruit/leaf (Yao et al. 2011)		
Scopoletin		Leaf/root bark (Zhou et al. 1996; Wei and Liang 2002; Hansel and Huang 1977)	

Table 1.3 (continued)

Compound name	L. barbarum	L. chinense	L. ruthenicum
Scopolin		Root bark (Wei and Liang 2002)	
Fabiatrin		Root bark (Wei and Liang 2002)	
Lignanoids			
(+)-Lyoniresinol 3α-O-β-D-glucopyranoside		Root bark (Han et al. 2002; Lee et al. 2005)	
Terpenoids			
L-Monomenthyl succinate	Fruit (Hiserodt et al. 2004)		
7, 8-Dehydro-3-hydroxy-β-ionone		Leaf (Sannai et al. 1984)	
L-Monomenthylglutarate		Fruit (Hiserodt et al. 2004)	
L-Dimenthylglutarate		Fruit (Hiserodt et al. 2004)	
A monoterpene glycoside		Root bark (Yahara et al. 1993)	
(−) -1, 2-Didehydro-α-cyperone		Fruit (Sannai et al. 1982)	
Solavetivone		Fruit (Sannai et al. 1982)	
Lyciumoside I		Leaf/stem/root bark (Yao et al. 2011)	
Lyciumoside II		Leaf/root bark (Yao et al. 2011)	
Lyciumoside III		Leaf/ root bark (Yao et al. 2011)	
Lyciumoside IV		Leaf (Yao et al. 2011)	
Lyciumoside V		Leaf (Yao et al. 2011)	
Lyciumoside VI		Leaf (Yao et al. 2011)	
Lyciumoside VII		Leaf (Yao et al. 2011)	
Lyciumoside VIII		Leaf (Yao et al. 2011)	
Lyciumoside IX		Leaf (Yao et al. 2011)	
Sugiol		Root bark (Noguchi et al. 1985)	
Sterols, steroids, and their derivatives			
B-sitosterol	Flower/fruit/stem (Yao et al. 2011)	Fruit/seed/root/root bark (Yao et al. 2011)	
Lanosterol	Flower (Harsh and Nag 1981)	Seed(Yao et al. 2011)	
β-Sitosterolβ-D-glucopyranoside	Fruit(Xie et al. 2001; Wang et al. 1998)	Root bark/leaf (Yao et al. 2011)	

Table 1.3 (continued)

Compound name	L. barbarum	L. chinense	L. ruthenicum
Diosgenin	Flower/stem (Harsh and Nag 1981)		
lycioside A	Seed (Wang et al. 2011)		
lycioside B	Seed (Wang et al. 2011)		
Campesterol		Fruit/seed/root (Yao et al. 2011)	
Cholesterol		Fruit/seed/root (Yao et al. 2011)	
24-Methylcholesterol		Seed (Itoh et al. 1977a)	
24-Ethylcholesterol		Seed (Itoh et al. 1977a)	
Stigmasterol		Fruit/seed/root (Yao et al. 2011)	
24-Methylcholesta-5, 24-dien-3β-ol		Seed (Itoh et al. 1977a)	
28-Isofucosterol		Seed (Itoh et al. 1977a)	
24-Methylidenecholesterol		Seed (Yao et al. 2011)	
24-Ethylcholesta-5, 24-dien-3β-ol		Seed (Yao et al. 2011)	
Cholestan-3β-ol		Seed (Itoh et al. 1977a)	
24-Methylcholestan-3β-ol		Seed (Itoh et al. 1977a)	
24-Ethylcholestan-3β-ol		Seed (Itoh et al. 1977a)	
Cholest-7-en-3β-ol		Seed (Yao et al. 2011)	
Cycloartanol		Seed (Itoh et al. 1977b)	
Cycloartenol		Seed (Yao et al. 2011)	
24-Methylidenecycloartanol		Seed (Yao et al. 2011)	
31-Norcycloartenol		Seed (Yao et al. 2011)	
31-Norlanosterol		Seed (Yao et al. 2011)	
31-Norcycloartanol		Seed (Itoh et al. 1978)	
31-Norlanost-8-enol		Seed (Yao et al. 2011)	
31-Norlanost-9(11)-enol		Seed(Itoh et al. 1978)	
24-Methyl-31-norla-nost-9(11)-enol		Seed (Itoh et al. 1978)	
4α-Methylcholest-8-en-3β-ol		Seed(Yao et al. 2011)	
4α,24-Dimethylcholesta-7,24-dien-3β-ol		Seed (Itoh et al. 1978)	
24-Ethyl-4α-methylcholesta-7,24-dien-3β-ol		Seed (Itoh et al. 1978)	
Cholest-5-en-3β-ol		Fruit (Yao et al. 2011)	

Table 1.3 (continued)

Compound name	*L. barbarum*	*L. chinense*	*L. ruthenicum*
24-Methylcholest-5-en-3β-ol		Fruit (Yao et al. 2011)	
24-Ethylcholesta-5,22-dien-3β-ol		Fruit (Yao et al. 2011)	
24-Ethylcholest-5-en-3β-ol		Fruit (Yao et al. 2011)	
24-Ethylidenecholest-5-en-3β-ol		Fruit (Yao et al. 2011)	
24-Ethylidenecholest-7-en-3β-ol		Fruit (Yao et al. 2011)	
Lophenol		Seed (Yao et al. 2011)	
24-Ethyllophenol		Seed (Itoh et al. 1978)	
24-Methyllophenol		Seed (Itoh et al. 1978)	
Gramisterol		Seed (Yao et al. 2011)	
Citrostadienol		Seed (Yao et al. 2011)	
Cycloeucalenol		Seed (Yao et al. 2011)	
Obtusifoliol		Seed (Yao et al. 2011)	
Lanost-8-enol		Seed (Yao et al. 2011)	
β-Amyrin		Seed (Yao et al. 2011)	
24-Methylidenelanost-8-en-3β-ol		Seed (Yao et al. 2011)	
Lupeol		Seed (Yao et al. 2011)	
5α-Stigmastane-3,6-dione		Root bark (Noguchi et al. 1985)	
β-Sitosterol 3-O-(6-palmitoyl-β-D-glucopyranoside)		Fruit (Jung et al. 2005)	
β-Sitosterol 3-O-(6-stearoyl-β- D-glucopyranoside)		Fruit (Jung et al. 2005)	
A furostanol glycoside		Root bark (Yahara et al. 1993)	
Lyciumsubstanz A		Leaf (Hansel et al. 1975; Hansel and Huang 1977)	
Lyciumsubstanz B		Leaf (Hansel and Huang 1977)	
Diosgenin		(Yao et al. 2011)	
Organic acids and their derivatives			
Linoleic acid	Fruit (Yao et al. 2011)	Fruit/root bark (Yao et al. 2011)	Fruit (Yao et al. 2011)
Oleic acid	Fruit (Wang et al. 1998)	Root (Yao et al. 2011)	Fruit (Yao et al. 2011)

Table 1.3 (continued)

Compound name	*L. barbarum*	*L. chinense*	*L. ruthenicum*
Melissic acid	Fruit (Yao et al. 2011)	Fruit/root bark (Yao et al. 2011)	Fruit (Yao et al. 2011)
Myristic acid	Fruit (Altintas et al. 2006)		
Palmitic acid	Fruit (Yao et al. 2011)	Root (Yao et al. 2011)	Fruit (Yao et al. 2011)
Caffeic acid	Fruit (Inbaraj et al. 2010)		
Ferulic acid	Fruit (Inbaraj et al. 2010)		
Vanillic acid	Leaf (Yao et al. 2011)	Leaf/root bark (Hansel and Huang 1977)	
Salicylic acid	Leaf (Yao et al. 2011)		
p-Coumaric acid	Fruit (Xie et al. 2001; Wang et al. 1998)		
Chlorogenic acid	Fruit (Dong et al. 2011b)		
Stearic acid		Root (Yao et al. 2011)	
Pentanoic acid		Fruit (Lee et al. 2010)	
Lactic acid		Leaf (Kim et al. 1997c)	
α-Dimorphecolic acid		Root bark (Yao et al. 2011)	
(9S, 10E, 12Z, 15Z)-9-Hydroxyoctadeca-10, 12, 15-trienoic acid		Root bark (Yao et al. 2011)	
Oxalic acid		Leaf (Kim et al. 1997c)	
Malonic acid		Leaf (Kim et al. 1997c)	
Malic acid		Leaf (Kim et al. 1997c)	
Succinic acid		Leaf (Kim et al. 1997c)	
Fumaric acid		Leaf (Kim et al. 1997c)	
Citric acid		Leaf (Kim et al. 1997c)	
Cinnamic acid		Root bark (Noguchi et al. 1985)	
Protocatechuic acid		Fruit (Qian et al. 2004)	
Chlorogenic acid		Leaf/fruit (Yao et al. 2011)	
β-D-Glucopyranosylsyringate		Root bark (Wei and Liang 2003)	
Stearylferulate		Root bark (Zhou et al. 1996)	
Anthocyanins			
Petundin-3-O-galacto-side-5-O-glucoside			Fruit (Zheng et al. 2011)

Table 1.3 (continued)

Compound name	*L. barbarum*	*L. chinense*	*L. ruthenicum*
Petundin-3-*O*-gluco-side-5-*O*-glucoside			Fruit (Zheng et al. 2011)
Delphinidin-3-*O*-rutinoside (cis-p-coumaroyl)-5-*O*-glucoside			Fruit (Zheng et al. 2011)
Delphinidin-3-*O*-rutinoside (trans-p-coumaroyl)-5-*O*-glucoside			Fruit (Zheng et al. 2011)
Petunidin-3-*O*-rutinoside (caffeoyl)-5-*O*-glucoside			Fruit (Zheng et al. 2011)
Pentunidin-3-*O*-rutinoside (cis-p-coumaroyl)-5-*O*-glucoside			Fruit (Zheng et al. 2011)
Pentunidin-3-*O*-rutinoside (trans-p-coumaroyl)-5-*O*-glucoside			Fruit (Zheng et al. 2011)
Pentunidin-3-*O*-glucoside (maloyl)-5-*O*-glucoside			Fruit (Zheng et al. 2011)
Pentunidin-3-*O*-glucoside (feruloyl)-5-*O*-glucoside			Fruit (Zheng et al. 2011)
Malvidin-3-*O*-rutinoside (cis-p-coumaroyl)-5-*O*-glucoside			Fruit (Zheng et al. 2011)
Glycolipids			
(2*S*) -1-*O*-Palmitoyl-2-*O*-linolenoyl-3-*O*-[α-D-galactopyranosyl- (1″-6′) - (3″-*O*-linolenoyl) -β-D-galactopyranosyl]-glycerol	Fruit (Gao et al. 2008)		
(2*S*) -1-*O*-Palmitoyl-2-*O*-linolenoyl-3-*O*-[α-D-galactopyranosyl- (1″-6′) - (3″-*O*-linolenoyl) -β-D-galactopyranosyl]-glycerol	Fruit (Gao et al. 2008)		
(2*S*) -1-*O*-Palmitoyl-2-*O*-linolenoyl-3-*O*-[α-D-galactopyranosyl- (1″-6′) - (3″-*O*-palmitoyl) -β-D-galactopyranosyl]-glycerol	Fruit (Gao et al. 2008)		
(2*S*) -1-*O*-Palmitoyl-2-*O*-linolenoyl-3-*O*-[α-D-galactopyranosyl- (1″-6′) - (3″-*O*-palmitoyl) -β-D-galactopyranosyl]-glycerol	Fruit (Gao et al. 2008)		

Table 1.3 (continued)

Compound name	*L. barbarum*	*L. chinense*	*L. ruthenicum*
(2*S*) -1-*O*-palmitoyl-2-*O*-palmitoyl-3-*O*-[-α-D-galactopyranosyl)-(1″-6′) - (3″-*O*-palmitoyl)-β-D-galactopyranosyl]glycerol	Fruit (Gao et al. 2008)		
(2*S*) -1-*O*-palmitoyl-2-*O*-palmitoyl-3-*O*-[-α-D-galactopyranosyl) - (1″-6′) -β-D-galactopyranosyl] glycerol	Fruit (Gao et al. 2008)		
(2*S*) -1-*O*-linolenoyl-2-*O*-linolenoyl-3-*O*-[α-D-galactopyranosyl- (1″-6′) -β-D-galactopyranosyl] glycerol	Fruit (Gao et al. 2008)		
(2*S*) -1-*O*-linolenoyl-2-*O*-linolenoyl-3-*O*-[α-D-galactopyranosyl- (1″-6′) -β-D-galactopyranosyl] glycerol	Fruit (Gao et al. 2008)		
(2*S*) -1-*O*-palmitoyl-2-*O*-linolenoyl-3-*O*-[α-D-galactopyranosyl- (1″-6′) -β-D-galactopyranosyl] glycerol	Fruit (Gao et al. 2008)		
(2*S*) -1-*O*-palmitoyl-2-*O*-linoleoyl-3-*O*-[α-D-galactopyranosyl- (1″-6′) -β-D-galactopyranosyl] glycerol	Fruit (Gao et al. 2008)		
(2*S*) -1-*O*-palmitoyl-2-*O*-oleoyl-3-*O*-[α-D-galactopyranosyl- (1″-6′) -β-D-galactopyranosyl] glycerol	Fruit (Gao et al. 2008)		
(2*S*) -1-*O*-Stearoyl-2-*O*-linoleoyl-3-*O*-[α-D-galactopyranosyl- (1″-6′) -β-D-galactopyranosyl] glycerol	Fruit (Gao et al. 2008)		
(2*S*) -1-*O*-palmitoyl-2-*O*-linolenoyl-3-*O*-β-D-galactopyranosylglycerol	Fruit (Gao et al. 2008)		
(2*S*) -1-*O*-palmitoyl-2-*O*-linoleoyl-3-*O*-β-D-galactopyranosylglycerol	Fruit (Gao et al. 2008)		

Table 1.3 (continued)

Compound name	L. barbarum	L. chinense	L. ruthenicum
(2S) -1-O-palmitoyl-2-O-oleoyl-3-O-β-D-galactopyranosylglycerol	Fruit (Gao et al. 2008)		
1-O-β-D-Galactopyranosyl-2, 3-O-bis-[(9Z, 12Z, 15Z)-octadeca-9, 12, 15-trienoyl] glycerol		Fruit (Jung et al. 2005)	
1-O-β-D-Galactopyranosyl-2-O-[(9Z, 12Z)-octadeca-9,12-dienoyl]-3-O-[(9Z, 12Z, 15Z-octadeca-9, 12, 15-trienoyl]glycerol		Fruit (Jung et al. 2005)	
α-D-glucuronopyranosyl (2→1′) -α-D -glucurono-pyranosyl (2′ →1″) -α-D-glucopyranosyl-2″-n-octadec-9‴-enoate		Fruit (Chung et al. 2013)	
abd-3β, 9β-diol-3α-D-glucopyranosyl- (2a →1b) -α-D-glucopyranosyl- (2b→1c) -α-D-glucopyranosyl- (2c→1d) -α-D-arabinofuranosyl-2d-p-hydroxybenzoate		Fruit(Chung et al. 2013)	

glucopyranoside was isolated from the root bark of *L. chinense* (Han et al. 2002; Lee et al. 2005). Two compounds were newly identified from the acetone extract of dry Goji berry, 3-(3-hydroxy-4-methoxyphenyl)-N-[2-(4-methoxyphenyl)ethyl]-(2E)-Propenamide, and 3-(4-hydroxy-3-methoxyphenyl)-N-[2-(4-hydroxyphenyl) ethyl]-2-Propenamide (Personal communication with Dr. Minghua Qiu).

1.5 Comparative Analysis of the Amount of Polysaccharide and Flavonoids in Leaves of Wild *L. Chinense* and *L. Barbarum* Plants

Considering leaves of both *L. chinense* and *L. barbarum* have been consuming as functional vegetable and functional tea, total amount of polysaccharide and flavo-noids were compared among leaves collected from wild plants (Yan et al. 2010; Dong et al 2011a, b and unpublished data). Six populations of the wild *L. barbarum* plants from Northwestern provinces of China, the polysaccharide content of the population from Zhongning (Ningxia province) (11.32 mg/g dry weight) was the highest while that from Dunhuang (Gansu province) (5.66 mg/g dry weight) was the lowest. Among the five populations of the wild *L. chinense* plants collected in the central provinces of China, the polysaccharide amount of the population from Yuncheng (Shanxi province) (12.14 mg/g) was the highest while that from Xixia (Henan province) was the lowest.

Young leaves of wild *L. chinense* has a significantly higher amount of polysaccharide than the wild *L. barbarum* plants (10.72 vs. 7.86 mg/g). It was reported that the divergence of the polysaccharide amount among different populations of each species was not significant, but the divergence among the individuals within each population was more significant, which indicates that such divergence may be due to the genetic divergence among individuals rather than the environmental differences (Yan et al. 2011). Therefore, we may obtain certain excellent germplasm of the high level of polysaccharide and total flavonoids in leaf through the germplasm screening.

Yan et al. (2010) also reported that among the six populations of the wild *L. barbarum* plants, the leaf total flavonoids amount of the population from Zhongning, Ningxia province (5.64 mg/g) was the highest while that from Nuomuhong, Qinghai province (2.87 mg/g) was the lowest. Additionally, it was shown that the interpopulation divergence was not significant, while the divergence within these populations was significant. For example, the highest amount of one sample is around 11-folds more than the lowest one of other sample in the Zhongning population. Moreover, the overall divergence for the leaf total flavonoids between wild *L. chinense* (3.998 mg/g) and *L. barbarum* (4.07 mg/g) was not significant.

The subsequent High-performance liquid chromatography (HPLC) analysis of the leaf total flavonoids revealed that although the divergence for the leaf total flavonoids between the two species was not significant, the composition and amount of the constituents were dramatically divergent (Fig. 1.2, unpublished data). Total flavonoids in the young leaves of these two species share the constituents like rutin, chlorogenic acid, kaempfero, quercetin, and apigenin-7-O- (6′-O-acetyl) glucose-rhamnose. Besides, there was another obvious peak for an unknown compound for the sample of the wild *L. barbarum* leaves which was however undetected for the wild *L. chinense* leaves [unpublished data in Wang's lab].

The Goji leaf is called "Tianjing grass" in the book *Compendium of Materia Medica* (*Ben Cao Gang Mu*) and serves as a good medicinal herb, functional vegetable, and functional tea. So, it is promising to develop the Goji leaf for both medicine and food uses. The Goji leaves from the cultivated and wild *L. chinense* and *L. barbarum* plants currently possess a large market share. However, the cultivation of the *L. barbarum* plants is limited in the northwest China, while *L. chinense* are widely distributed and consumed in the southern and central China. Besides, the leaves of *L. chinense* can be more nutritional than *L. barbarum* plants. Therefore, it is important to breed good Goji cultivars of good nutrition and medicinal use for various regions. To achieve this goal, we need to have more in-depth analyzes of the amount and composition of active constituents in the leaves of various species cultivated under different conditions.

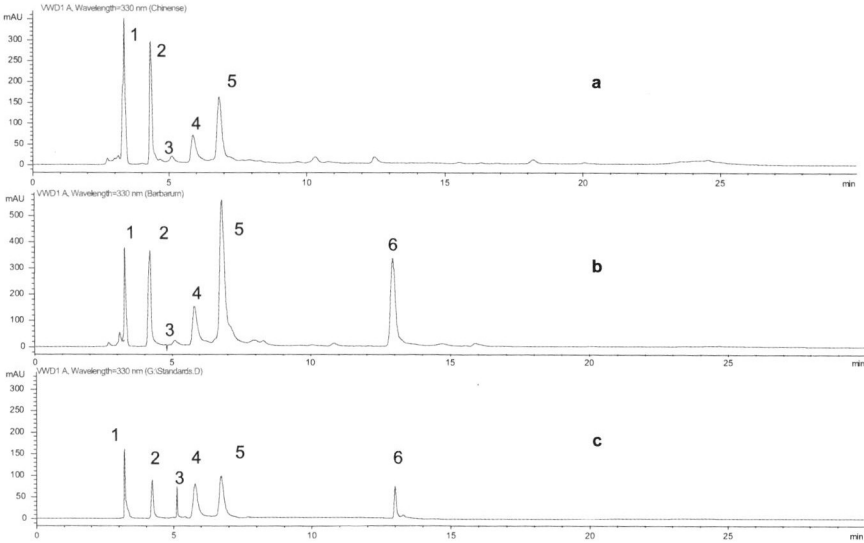

Fig. 1.2 Comparison of the typical HPLC of the total flavonoids extracted from the wild *L. chinense* and *L. barbarum* leaves with those of the standards **a** *L. chinense*, **b** *L. barbarum*, **c** standards; *1* rutin; *2* chlorogenic acid; *3* quercetin; *4* kaempfero; *5* apigenin-7-O- (6'-O-acetyl) glucose-rhamnose; *6* unknown compound). (Dong et al. 2011a)

1.6   Analysis of the Major Constituents in *L. barbarum* Fruits (Goji Berry) Cultivated in Various Regions

In order to compare the Goji berry quality from various production areas, the dry Goji berry of "Ningqi 1" cultivar were collected in the typical cultivation sites located in eight regions including Jinghe and Qitai in Xinjiang, Shahai in Inner Mongolia, Zhouta Village in Zhongning in Ningxia, Nuomuhong and Geermu in Qinghai, Dazi in Tibet, and Jvlu in Hebei (Dong et al. 2012). To avoid variation derived from the different drying process, fresh fruits were desiccated using the silica gels till their weight remaining constant. They were then subject to phytochemical analysis following method published by Dong et al. 2012.

Phytochemicals and fruit characters are summarized in Table 1.4, which is adjusted from Dong (Dong et al. 2014). Based on the 100-fruit weight, Goji berry from Nuomuhong is the heaviest with 19.71 g for 100 complete fruits and 16.95 g for 100 fruit pulp. In ancient traditional Chinese medicine system, Zhongning has been authenticated as the genuine producing area where Goji has the highest medicinal quality. Zhongning Goji was reported to have the highest content of flavonoids (2.97%) and the highest antioxidant activity, which indicated that content of flavonoids in Goji berry can accurately reflect the antioxidant ability and confirm that Zhongning is the genuine producing area for polyphenols. Goji berry produced in Jvlu, Hebei Province, has the highest carotenoids amount which diverges significantly from other regions. This result is consistent with the brand characteristics of

Table 1.4 Characteristic parameters of the cultivated *L. barbarum* dry fruits produced in various regions. (Source: Dong et al. 2012)

Location	Total flavonoids (%)	Total carotenoids (%)	Polysaccharide (%)	100-berry-weight (g)	1000-seed-weight (g)	Seed number per berry	Flesh/seed ratio (g/g)
Zhongning	2.94 A*	0.10 C	5.37 C	12.18	0.59	30.0	5.93
Shahai	2.71 B	0.15 B	6.96 B	13.74	0.94	25.0	4.84
Jvlu	1.58 E	0.16 A	7.01 B	8.95	0.90	20.0	3.94
Jinghe	2.45 D	0.07 D	8.90 A	13.31	1.23	12.0	8.04
Qitai	2.47 D	0.09 C	7.32 B	7.98	0.54	17.0	7.71
Nuomuhong	2.56 C	0.05 E	7.23 B	19.71	0.73	38.0	6.12
Dazi	2.63 C	0.06 E	7.42 B	17.98	0.70	36.5	6.02
Geermu	2.64 C	0.05 E	7.36 B	18.53	0.66	33.2	7.49
RSD, %	3.56–9.21	1.85–7.41	3.71–5.87	2.06–3.15	1.17–2.1	0.75–1.12	2.6–3.5

The numbers in the table were the averages of five repeats; the uppercased letters after the numbers represented the statistics for multiple comparison; different letters indicated significant divergence ($p < 0.01$)

the Goji fruits produced in Jvlu. In the long history for consuming Goji, the Jvlu Goji berry has been renowned as "blooded Goji berry" and their deep red color may be attributed to the high carotenoids amount. The fruits produced in Shahai possessed the second highest amount of carotenoids, followed by Zhongning and Qitai. As to the polysaccharide amount in Goji berry from various regions, it was found that Jinghe in Xinjiang possessed the highest polysaccharide level and Zhongning Goji had the lowest level. Interestingly, an obvious decreasing trend of polysaccharide content from west (Xinjiang) to east (Ningxia) is positively related to the temperature, indicating the areas of Northwestern China with relatively low temperature are good for polysaccharide accumulation, but not the best for carotenoid or flavonoid accumulation.

The Goji berry is not only a precious traditional Chinese medicine, but also a functional food with both the medicine and food uses. Goji plants have been widely cultivated with a long cultivation history. Zhongning County has been regarded in Chinese history as the region producing genuine Goji berries. There is a record in the famous book *Compendium of Materia Medica* (*Ben Cao Gang Mu*) that the Goji berries used as medicines in the whole country were all produced in the region around Zhongning. The cultivars in this region have been later introduced to the regions outside Zhongning, including Jvlu in Hebei Province, Shahai in Inner Mongolia, Nuomuhong in Qinghai Province, Jinghe in Xingjiang, and Guazhou in Gansu Province, which results in the expansion of the cultivation region for Goji fruit production. Due to the diversity of the environmental factors and the active constituents of Goji berry, optimal production regions for a specific Goji berry of various dominant constituents should be established based on various usage and active constituent composition. For example, the polysaccharide region (Jinghe in Xinjiang province), the flavonoid region (Zhongning in Ningxia province), the carotenoid region (Jvlu in Hebei province) and the regular regions should be clearly designated for various market share. This strategy will benefit the quality control of Goji berry products and scientifically choosing of raw materials for various usages. Further expansion of existing and potential production regions is of extreme significance for the development of Goji industry worldwide and contributing to the human health.

1.7 Summary and Future Research Direction

High level of genetic diversity has been discovered in wild *Lycium* resources. The molecular studies of Goji include identification of genes involved in the secondary metabolism, disease resistance, and pest resistance. New projects of genomics, transcriptomics, metabolomics have been initiated to study the evolution of whole genome and gene families, identify important genes, and promote the molecular breeding program. Currently, it is urgent to breed unique cultivars for various production regions and diversified products, including cultivars used specifically for juice, medicine, functional vegetable and tea, fresh fruit, and pure compound or essential oil extracts. Because the cultivation of Goji plants is highly labor consuming, with 1/3 of price accounting the labor cost (especially hand harvesting), new cultivars for mechanical harvesting are also under development.

With the development of Goji industry worldwide, the classical cultivation pattern for producing mainly dry fruits using air-dry method has gradually changed into the machine drying in order to retain the high level of bioactive components. We found that the polysaccharide amount of fresh fruits or fruits desiccated using silica gels was obviously higher than that of the fruits dried using the default method recommended by the Chinese Pharmacopoeia or the air-dry method. Different drying facilities using electricity, solar energy, or coals have been invented and used in various production regions.

The quality of Goji berry and phytochemical contents are affected by cultivars, environmental factors, cultivation patterns, and processing methods. The traditional quality standard of Goji berry just pays attention to the shape and size of dry fruits. Therefore, it is also urgent to establish a comprehensive quality standard considering both the morphological characteristics and phytochemical levels, which is more scientific and accurate in analyzing the Goji berry samples. Moreover, the active constituents in Goji berry and leaf and their efficacy await further study. In-depth knowledge will facilitate the development of new products or Goji-based medicines. Together, the breakthrough in basic research, breeding, and cultivation and processing technology is of ultimate significance to guide the efficient and fast development of Goji industry and the modernization of traditional Chinese medicines.

References

Adams M, Wiedenmann M, Tittel G, Bauer R. HPLC-MS trace analysis of atropine in *Lycium barbarum* berries. Phytochem Anal. 2006;17:279–83.

Altintas A, Kosar M, Kirimer N, Baser KHC, Demirci B. Composition of the essential oils of *Lycium barbarum* and *L. ruthenicum* fruits. Chem Nat Compd. 2006;42:24–5.

Amagase H, Farnswoeth NR. A review of botanical characteristics, phytochemistry, clinical relevance in efficacy and safety of *Lycium barbarum* fruit (Goji). Food Res Int. 2011;44:1702–17.

Asano N, Kato A, Miyauchi M, Kizu H, Tomimori T, Matsui K, Nash RJ, Molyneux RJ. Specific alpha-galactosidase inhibitors, N-methylcalystegines-structure/activity relationships of calystegines from *Lycium chinense*. Eur J Biochem. 1997;248:296–303.

Chen Z, Tan BKH, Chan SH. Activation of T lymphocytes by polysaccharide-protein complex from *Lycium barbarum* L. Int Immunopharmacol. 2008;8:1663–71.

Chin YW, Lim SW, Kim SH, Shin DY, Suh YG, Kim YB, Kim YC, Kim J. Hepatoprotective pyrrole derivatives of *Lycium chinense* fruits. Bioorg Med Chem Lett. 2003;13:79–81.

Chung IM, Ali M, Nagella P, Ahmad A. New glycosidic constituents from fruits of *Lycium chinense* and their antioxidant activities. Arabian J Chem. 2013. http://dx.doi.org/10.1016/j.arabjc.2013.1005.1020.

Dong J, Gao WS, Lu D, Wang Y. Simultaneous extraction and analysis of four polyphenols from leaves of *Lycium barbarum* L. J Food Biochem. 2011a;35:914–31.

Dong JZ, Gao WS, Lu DY, Wang Y. Simultaneous extraction and analysis of four polyphenols from leaves of *Lycium barbarum* L. J Food Biochem. 2011b;35:914–31.

Dong J, Wang S, Wang Y. Analysis on the main active components of *Lycium barbarum* fruits and related environmental factors. J Med Plants Res. 2012;6:2276–83.

Duan C, Qiao S, Wang N, Zhao Y, Qi C, Yao X. Studies on the active polysaccharides from *Lycium barbarum* L. Acta Pharm Sinica. 2001;36:196–9.

Funayama S, Yoshida K, Konno C, Hikino H. Validity of oriental medicines .21. Structure of kukoamine-a, a hypotensive principle of *Lycium chinense* root barks. Tetrahedron Lett. 1980;21:1355–6.

Gan L, Zhang SH, Yang XL, Xu HB. Immunomodulation and antitumor activity by a polysaccharide-protein complex from Lycium barbarum. Int Immunopharmacol. 2004;4:563–9.

Gao ZP, Ali Z, Khan IA. Glycerogalactolipids from the fruit of *Lycium barbarum*. Phytochemistry. 2008;69:2856–61.

Han BH, Park JH, Park MH, Han YN. Studies on the alkaloidal components of the fruits of *Lycium chinense*. Arch Pharm Res. 1985;8:249–52.

Han SH, Lee HH, Lee IS, Moon YH, Woo ER. A new phenolic amide from *Lycium chinense* Miller. Arch Pharm Res. 2002;25:433–7.

Hansel R, Huang JT. *Lycium chinense*-semiquantitative assay of with anolides. Arch Pharm. 1977;310:35–8.

Hansel R, Huang JT, Rosenberg D. Withanolides from *Lycium chinense*. Arch Pharm. 1975;308:653–4.

Harsh ML. Tropane alkaloids from *Lycium barbarum* Linn in vivo and in vitro. Curr Sci. 1989;58:817–8.

Harsh ML, Nag TN. Diosgenin and phytosterols from *Lycium barbarum* Linn. Curr Sci. 1981;50:235.

Hiserodt RD, Adedeji J, John TV, Dewis ML. Identification of monomenthyl succinate, monomenthyl glutarate, and dimenthyl glutarate in nature by high performance liquid chromatography-tandem mass spectrometry. J Agric Food Chem. 2004;52:3536–41.

Huang L, Lin Y, Tian G, Ji G. Isolation, purification and physico-chemical properties of immunoactive constituents from the fruit of *Lycium barbarum* L. Acta Pharm Sinica. 1998;33:512–6.

Huang LJ, Tian GY, Ji GZ. Structure elucidation of glycan of glycoconjugate lbgp3 isolated from the fruit of *Lycium barbarum* L. J Asian Nat Prod Res. 1999;1:259–67.

Inbaraj BS, Lu H, Kao TH, Chen BH. Simultaneous determination of phenolic acids and flavonoids in *Lycium barbarum* Linnaeus by HPLC-DAD-ESI-MS. J Pharm Biomed Anal. 2010;51:549–56.

Itoh T, Tamura T, Matsumoto T. 4-Desmethylsterols in seeds of Solanaceae. Steroids. 1977a;30:425–33.

Itoh T, Tamura T, Matsumoto T. Triterpene alcohols in seeds of Solanaceae. Phytochemistry. 1977b;16:1723–6.

Itoh T, Ishii T, Tamura T, Matsumoto T. 4 New and other 4-alpha-methylsterols in seeds of Solanaceae. Phytochemistry. 1978;17:971–7.

Jang YP, Lee YJ, Kim YC, Huh H. Production of a hepatoprotective cerebroside from suspension cultures of *Lycium chinense*. Plant Cell Rep. 1998;18:252–4.

Jung KW, Chin YW, Kim YC, Kim J. Potentially hepatoprotective glycolipid constituents of *Lycium chinense* fruits. Arch Pharm Res. 2005;28:1381–5.

Kim SY, Choi YH, Huh H, Kim J, Kim YC, Lee HS. New antihepatotoxic cerebroside from *Lycium chinense* fruits. J Nat Prod. 1997a;60:274–6.

Kim SY, Kim HP, Huh H, Kim YC. Antihepatotoxic zeaxanthins from the fruits of *Lycium chinense*. Arch Pharm Res. 1997b;20:529–32.

Kim SY, Lee KH, Chang KS, Bock JY, Jung MY. Taste and flavor compounds in box thorn (*Lycium chinense* Miller) leaves. Food Chem. 1997c;58:297–303.

Kim SY, Lee EJ, Kim HP, Lee HS, Kim YC. LCC, a cerebroside from *Lycium chinense*, protects primary cultured rat hepatocytes exposed to galactosamine. Phytother Res. 2000;14:448–51.

Le K, Chiu F, Ng K. Identification and quantification of antioxidants in *Fructus lycii*. Food Chem. 2007;105:353–63.

Lee DG, Park Y, Kim MR, Jung HJ, Seu YB, Hahm KS, Woo ER. Anti-fungal effects of phenolic amides isolated from the root bark of *Lycium chinense*. Biotechnol Lett. 2004;26:1125–30.

Lee DG, Jung HJ, Woo ER. Antimicrobial property of (+)-lyoniresinol-3 alpha-O-beta-D-gluco-pyranoside isolated from the root bark of *Lycium chinense* Miller against human pathogenic microorganisms. Arch Pharm Res. 2005;28:1031–6.

Lee HJ, Ahn HJ, Kang CS, Choi JC, Choi HJ, Lee KG, Kim JI, Kim HY. Naturally occurring pro-pionic acid in foods marketed in South Korea. Food Control. 2010;21:217–20.

Levin R, Bernardello G, Whiting C, Miller J. A new generic circumscription in tribe *Lycieae* (So-lanaceae). Taxon. 2011;60:681–90.

Liang B, Jin M, Liu H. Water-soluble polysaccharide from dried *Lycium barbarum* fruits: isola-tion, structural features and antioxidant activity. Carbohyd Polym. 2011;83:1947–51.

Liu H, Fan Y, Wang W, Liu N, Zhang H, Zhu Z, Liu A. Polysaccharides from *Lycium barbarum* leaves: isolation, characterization and splenocyte proliferation activity. Int J Biol Macromol. 2012;51:417–22.

Liu Y, Sun W, Zeng S, Huang W, Liu D, Hu W, Shen X, Wang Y. Virus-induced gene silencing in two novel functional plants, *Lycium barbarum* L. and *Lycium ruthenicum* Murr. Sci Hortic. 2014;170:267–74.

Lv X, Wang C, Cheng Y, Huang L, Wang Z. Isolation and structural characterization of a polysac-charide LRP4-A from *Lycium ruthenicum* Murr. Carbohydr Res. 2013;365:20–5.

Miean KH, Mohamed S. Flavonoid (myricetin, quercetin, kaempferol, luteolin, and apigenin) con-tent of edible tropical plants. J Agric Food Chem. 2001;49:3106–12.

Noguchi M, Mochida K, Shingu T, Kozuka M, Fujitani K. On the components of the chinese drug ti-ku-pi.1. Isolation and composition of lyciumamide, a new dipeptide. Chem Pharm Bull. 1984;32:3584–7.

Noguchi M, Mochida K, Shingu T, Fujitani K, Kozuka M. Studies on the constituents of chinese drug ti-ku-pi.2. Sugiol and 5-alpha-stigmastane-3,6-dione from the chinese drug ti-ku-pi (*lycii radicis*-cortex). J Nat Prod. 1985;48:342–3.

Peng X, Tian G. Structural characterization of the glycan part of glycoconjugate LbGp2 from *Lycium barbarum* L. Carbohydr Res. 2001;331:95–9.

Peng XM, Huang LJ, Qi CH, Zhang YX, Tian GY. Studies on chemistry and immuno-modulating mechanism of a glycoconjugate from *Lycium barbarum* L. Chin J Chem. 2001a;19:1190–7.

Peng XM, Qi CH, Tian GY, Zhang YX. Physico-chemical properties and bioactivities of a glyco-conjugate LbGp5B from *Lycium barbarum* L. Chin J Chem. 2001b;19:842–6.

Peng Q, Lv X, Xu Q, Li Y, Huang L, Du Y. Isolation and structural characterization of the polysac-charide LRGP1 from *Lycium ruthenicum*. Carbohydr Res. 2012a;90:95–101.

Peng Q, Song J, Lv X, Wang Z, Huang L, Du Y. Structural characterization of an arabinogalactan-protein from the fruits of *Lycium ruthenicum*. J Agric Food Chem. 2012b;60:9424–9.

Potterat O. Goji (*Lycium barbarum* and *L. Chinense*): phytochemistry, pharmacology and safety in the perspective of traditional uses and recent popularity. Planta Med. 2010;76:7–19.

Qian JY, Liu D, Huang AG. The efficiency of flavonoids in polar extracts of *Lycium chinense* Mill fruits as free radical scavenger. Food Chem. 2004;87:283–8.

Qin X, Yamauchi R, Aizawa K, Inakuma T, Kato K. Isolation and characterization of arabinogalac-tan-protein from the fruit of *Lycium chinense* Mill. J Appl Glycosci. 2000;47:155–61.

Qin XM, Yamauchi R, Aizawa K, Inakuma T, Kato K. Structural features of arabinogalactan-proteins from the fruit of *Lycium chinense* Mill. Carbohydr Res. 2001;333:79–85.

Sannai A, Fujimori T, Kato K. Isolation of (-)-1,2-dehydro-alpha-cyperone and solavetivone from *Lycium chinense*. Phytochemistry. 1982;21:2986–7.

Sannai A, Fujimori T, Kato K. Neutral volatile components of kukoshi (*Lycium chinense* M). Agric Biol Chem. 1983;47:2397–9.

Sannai A, Fujimori T, Uegaki R, Akaki T. Isolation of 3-hydroxy-7,8-dehydro-beta-ionone from *Lycium chinense* M. Agric Biol Chem. 1984;48:1629–30.

Wang Q, Qiu Y, He SP, Chen YY. Chemical constituents of the fruit of *Lycium barbarum* L. J Chin Pharm Sci. 1998;7:218–20.

Wang K, Sasaki T, Li W, Li Q, Asada Y, Kato H, Koike K. Two novel steroidal alkaloid glycosides from the seeds of *Lycium barbarum*. Chem Biodivers. 2011;8:2277–84.

Wei XL, Liang JY. Chemical study on the root barks of *Lycium chinese* Mill. J China Pharm Univ. 2002;33:271–3.

Wei XL, Liang JY. Chemical studies on root bark of *Lycium chinense*. Chin Tradit Herb Drugs. 2003;34:580–1.

Xie C, Xu LZ, Li XM, Li KM, Zhao BH, Yang SL. Studieds on chemical constituents in fruit of *Lycium barbarum* L. China J Chin Mater Med. 2001;26:323–4.

Yahara S, Shigeyama C, Nohara T, Okuda H, Wakamatsu K, Yasuhara T. Structures of anti-ace and anti-renin peptides from lycii-radicis cortex. Tetrahedron Lett. 1989;30:6041–2.

Yahara S, Shigeyama C, Ura T, Wakamatsu K, Yasuhara T, Nohara T. Cyclic-peptides, acyclic diterpene glycosides and other compounds from *Lycium chinense* mill. Chem Pharm Bull. 1993;41:703–9.

Yan X, Dong J, Wang Y. Comparison studies of main active compounds in young leaves of *L. barbarum* and *L. chinense*. Food Sci. 2010;31:29–32.

Yao X, Peng Y, Xu L-J, Li L, Wu Q-L, Xiao P-G. Phytochemical and biological studies of *Lycium* medicinal plants. Chem Biodiv. 2011;8:976–1010.

Zeng S, Wu M, Zou C, Liu X, Shen X, Hayward A, Liu C, Wang Y. Comparative analysis of anthocyanin biosynthesis during fruit development in two *Lycium* species. Physiol Plant. 2014;150:505–16.

Zhang Z, Lu A, William GDA. Flora of China, Vol. 17. Beijing:Science;1994.300–32.

Zhao CJ, He YQ, Li RZ, Cui GH. Chemistry and pharmacological activity of peptidoglycan from *Lycium barbarum* L. Chin Chem Lett. 1996;7:1009–10.

Zhao CJ, Li RZ, He YQ, Cui GH. Studies on the chemistry of Gouqi polysaccharides. J Beijing Med Univ. 1997;29:231–40.

Zheng J, Ding C, Wang L, Li G, Shi J, Li H, Wang H, Suo Y. Anthocyanins composition and antioxidant activity of wild *Lycium ruthenicum* Murr. from Qinghai-Tibet plateau. Food Chem. 2011;126:859–65.

Zhou XW, Xu GJ, Wang Q. Studies on the chemical constituents in roots of *Lycium chinense* Mill. China J Chin Mater Med. 1996;21:675–6.

Zou YH. Flavones in leaves of *Lycium Chinense*. J Instrum Anal. 2002;21:76–8.

Chapter 2
Immunoregulation and *Lycium Barbarum*

Xiaorui Zhang, Wenxia Zhou and Yongxiang Zhang

Abstract Widespread pharmacology research has revealed that immune regulation is the main biological effect of *Lycium barbarum*, and the material bases to elicit the effect of *L. barbarum* on the immune system are polysaccharide and glycoprotein complexes. Studies have found that the effect of *L. barbarum* polysaccharide on the activation of macrophages and dendritic cells are important in participating in the immune response. In addition, it was found that TLR4/2 may be closely related to the immunoregulatory activity of *L. barbarum* polysaccharide. Further, the TLR4/2-activated signaling pathways lead to activation of phosphoinositide-3-kinase (PI3K) and LKB1, leading to activation of the mitogen-activated protein kinase (MAPK), extracellular signal-regulated kinase (ERK) and nuclear factor-κB (NF-κB), p53, C-Jun and AP-1. These pathways lead to induction of gene transcription. *L. barbarum* can induce production of a variety of cytokines, such as the anti-inflammatory factor IL-10, proinflammatory cytokines IL-1ß and IL-6, chemokines IL-8, antitumor factors cytokine TNF-α, antiviral factor IFN-γ, TGF-ß1, and lymphocyte activators IL-2 and IL-4. *L. barbarum* polysaccharide could also upregulate CD40, CD80, CD86, and major histocompatibility complex (MHC) class II molecules to various extents, and enhance antibody titers. Ultimately, activation of these transcription pathways induces expression of pro-inflammatory cytokines and immune regulation, survival, and proliferation.

Keywords *Lycium barbarum* polysaccharide · Polysaccharide-protein complex · Lymphocytes · Macrophage · TLR4 · NF-κB · AP-1

X. Zhang (✉) · W. Zhou · Y. Zhang
State Key Laboratory of Toxicology and Medical Countermeasures, Beijing Institute of Pharmacology and Toxicology, Beijing 100850, People's Republic of China
e-mail: zhangx_r@126.com

W. Zhou
e-mail: zhouwx@bmi.ac.cn

Y. Zhang
e-mail: zhangyx@bmi.ac.cn

2.1 Introduction

Widespread pharmacology research has revealed that immune regulation is the main biological effect of *Lycium barbarum*. Research indicates that the main material bases to elicit the effect of *L. barbarum* on the immune system are polysaccharide and glycoprotein complexes, in addition to other substances such as volatile oils, vitamins, and so on, which also exert certain immune effects (Huang et al. 2001; Tian et al. 1995; Zhao et al. 1996). At present, the immune regulation mechanism of *L. barbarum* is known to be mainly related to the following aspects. First, *L. barbarum* polysaccharides or glycoprotein compounds can activate macrophages, dendritic cells, and T cells to achieve cellular and humoral immune responses (Chen et al. 2008b; Nan et al. 2012). Second, a study found that the activity of *L. barbarum* polysaccharide (LBP) is related to TLR4/2 and that LBP can influence the PI3K/Akt/FoxO1, LKB1/AMPK, JNK/c-Jun, MEK/ERK, and nuclear factor kappa B pathways (Xiao et al. 2013). However, thus far, the exact molecular mechanisms underlying the effects of LBP are still unclear, especially the interaction of the polysaccharide with its receptors and its binding molecules. *L. barbarum* has various biological activities, such as modulation of blood vessels and blood flow, antitumor activity, prevention of neurodegeneration from Alzheimer's disease, stimulation of neurogenesis to improve sexual function, neuroprotective effects in ischemic stroke, skin-related effects from oral and topical preparations and improvement of vision and glaucoma (Chan et al. 2007; Ho et al. 2010). Indeed, wild-spectrum for the effects of *L. barbarum* is not surprising. Under physiological conditions, the immune system itself is involved in every system of the body via cytokines and the antibody network, including systems such as the nervous system. Under pathophysiological conditions, all diseases are directly or indirectly associated with the immune responses. To better outline the effects of *L. barbarum* on the immune system and the underlying mechanisms, this section will focus on our current understanding of the target cells of *L. barbarum*, activity-related receptors of LBP, *L. barbarum*-associated intracellular signal transduction, and effects of *L. barbarum* on the production of cytokines, antibodies, and some other functional molecules in the immune system.

2.2 Material Basis of Immunomodulation

The *L. barbarum* phytochemical diversity includes polysaccharides, carotenoids, flavonoids, alkaloids (Wang et al. 2011), peptides (Yuan et al. 2008), sterols (Park et al. 2012), organic acids (Inbaraj et al. 2010; Dong et al. 2013), essential oils (Altintas et al. 2006), glycolipids (Gao et al. 2008), polyphenols (Dong et al. 2011) and so on (Wu et al. 2012; Wang et al. 2010a). Presently, only polysaccharides and polysaccharide-protein complex have been reported to possess immune regulation activity.

2.2.1   *Polysaccharides*

Polysaccharides are the main active components of *L. barbarum* and also form the basis of its immunoregulatory activity. In previous studies, the calculated extraction yield of polysaccharides from the fruit of *L. barbarum* was approximately 9.43~13% (Archer and Mench 2014; Li et al. 2007), and the yield of polysaccharides from the leaves of *L. barbarum* was 16.2% (Liu et al. 2012). *L. barbarum* polysaccharides (LBPs) are known to have a variety of immunomodulatory functions (Zhu et al. 2007). LBPs are capable of promoting both the phenotypic and functional maturation of murine bone marrow-derived dendritic cells (BMDC) in vitro. LBPs can also increase peripheral WBC counts of chemotherapy-induced myelosuppressive mice to some extent, but there are no significant differences when compared with controls, and LBPs can obviously stimulate human PBMCs to produce G-CSF (Amagase et al. 2008). Moreover, it has been reported that LBPs can be used in compensating for the decline in total antioxidant capacity (TAOC), immune functions, and the activities of antioxidant enzymes and thereby reducing the risks of lipid peroxidation accelerated by age-induced free radicals (Li et al. 2007). These studies indicate that polysaccharides are one of the main immunoregulatory components of *L. barbarum*.

2.2.2   *Polysaccharide-Protein Complex*

Five homogenous polysaccharide-protein complexes have been obtained from *L. barbarum* fractions, designated LBPF1, LBPF2, LBPF3, LBPF4, and LBPF5. The carbohydrate contents of LBPF1–5 were 48.2, 30.5, 34.5, 20.3, and 23.5%, respectively, as measured by phenol-sulfuric acid assays using glucose as the standard. The protein contents were 1.2, 4.8, 4.1, 13.7, and 17.3%, respectively, as measured by the Bradford method (Chen et al. 2008a). Research has found that LBPF1–5 can activate T lymphocytes and enhance the Th1 response, and polysaccharide–protein complex-treated DCs displayed enhanced Th1 and Th2 responses in vitro and in vivo (Chen et al. 2008b, 2009a). Another fraction, LBP3p, at 10 mg/kg upregulate phagocytic activities of macrophages, altered the antibody type secreted by spleen cells, increased proliferation of spleen lymphocytes, increased CTL activity, increased the mRNA expression level for IL-2 and reduced the lipid peroxidation in S180-bearing mice (Gan et al. 2004). A polysaccharide–protein complex isolated from *L. barbarum* can also enhance innate immunity by activating macrophages (Chen et al. 2009b). These reports indicate that polysaccharide-protein complex is another important immunoregulation component of *L. barbarum*.

2.2.3   Others

Polyphenols and essential oil have immunoregulatory functions. Several biological activities have been described for polyphenolic compounds, including a modulator effect on the immune system. The effects of these biologically active compounds on the immune system are associated with processes such as differentiation and activation of immune cells (Cuevas et al. 2013). The essential oil constituents from aromatic herbs and dietary plants include monoterpenes, sesquiterpenes, oxygenated monoterpenes, oxygenated sesquiterpenes and phenolics, among others. Various mechanisms, such as enhancement of immune function and surveillance, are responsible for their chemopreventive properties (Bhalla et al. 2013). Polyphenols and essential oil from *L. barbarum* have been reported. However, it is unclear of whether the polyphenols and the essential oil are the immune regulation active component of *Lycium barbarum*.

2.3   Immunomodulatory Mechanism

The immunostimulatory activity of *L. barbarum* was studied and described almost 13 years ago (Huang et al. 2001; Peng et al. 2001). Shortly after those reports, the effects of *L. barbarum* polysaccharide-protein complex (LBP3p) on the expression of interleukin-2 and TNF-α in human peripheral blood mononuclear cells were investigated by reverse transcription polymerase chain reaction (RT-PCR) and bioassay (Gan et al. 2003). Then, modulation of a polysaccharide-protein complex from *L. barbarum* (LBP3p) on the immune system in S180-bearing mice was investigated (Gan et al. 2004). The neuroprotective effects of *L. barbarum* in a rat chronic ocular hypertension model via immunomodulation of macrophages/microglia was observed (Ip et al. 2006). LBPs were then reported to regulate phenotypic and functional maturation of murine dendritic cells (Zhu et al. 2007). In 2008, the first report of immune modulation by a standardized *L. barbarum* fruit (Goji) juice in randomized, double-blind, placebo-controlled clinical studies was published (Amagase et al. 2008). After that, many studies observed the target cells of *L. barbarum*, including T lymphocytes, macrophages, and dendritic cells (Chen et al. 2008b, 2009a, b; Zhang et al. 2011). More recently, the molecular mechanism and activity-related receptors were further reported, including I kappa B phosphorylation, as well as NF-κB, p65, p50, TLR4, and TLR2 (Chen et al. 2012; Wu et al. 2012; Zhu et al. 2013; Zhang et al. 2014b). A summary of the cellular targets and signal transduction pathways of *L. barbarum* is provided in Table 2.1 and Fig. 2.1. As many studies have reported the action of *L. barbarum* in activating innate immunity, affecting adaptive immunity, and inducing humoral and cell-mediated immune responses to better understand the immunoregulatory mechanisms of *L. barbarum*, we will discuss the issue of the target cells of *L. barbarum*, the active related receptors for LBP, the *L. barbarum*-associated signal transduction molecules, and also the changes of cytokines, antibodies, and leukocyte differentiation antigens affected by *L. barbarum* or LBP.

Table 2.1 Effect of *L. barbarum* on immune target cell

Dendritic Cell		T lymphocyte	
	Promotes: Maturation Antigen presenting		*Promotes:* Activation Proliferation
	Increases: IL-12p40 CD40 IL-12p70 CD80 CD11c CD86 I-A/I-E		*Increases:* IL-2 IFN-gama NFAT AP-1 CD25
Macrophage	*Promotes:* Maturation Antigen presenting	NK cell	*Promotes:* Cytotoxicity
	Increases: IL-12p40 CD40 TNF-alpha CD80 IL-1 beta CD86 Nitric oxide CD11c NF-κB I-A/I-E AP-1		*Increases:* IFN-γ perforin *Inhibits:* Apoptosis Necrosis

2.3.1 Target Cells

2.3.1.1 Dendritic Cells

Dendritic cells (DCs) represent a heterogeneous population of antigen-presenting cells that initiate the primary immune response (Banchereau et al. 2000). These cells take up antigens in peripheral tissues and migrate to secondary lymphoid organs where they become mature and competent in presenting antigens to T cells, thus initiating antigen-specific immune responses or immunological tolerance (Guermonprez et al. 2002). DC immunogenicity correlates with the DC functionally mature state, which is characterized by high-level expression of MHC and T cell costimulatory molecules, acute decreases in antigen uptake, and the ability to present antigens captured in the periphery to T cells (Wilson and Villadangos 2005). DC maturation can be induced by microbial products (such as LPS) or inflammatory cytokines (such as TNF) (Winzler et al. 1997). Although these mediators are potent stimuli of DC maturation, they are toxic and have limited applications. In this regard, as biological response modifiers (BRMs), polysaccharides are able to induce DC maturation and immunogenicity.

LBPs are known to exhibit immunomodulatory functions, including activation of B cells and natural killer (NK) cells. However, little is known about the immunomodulatory effects of LBPs on DC. The effects of LBPs on the phenotypic and

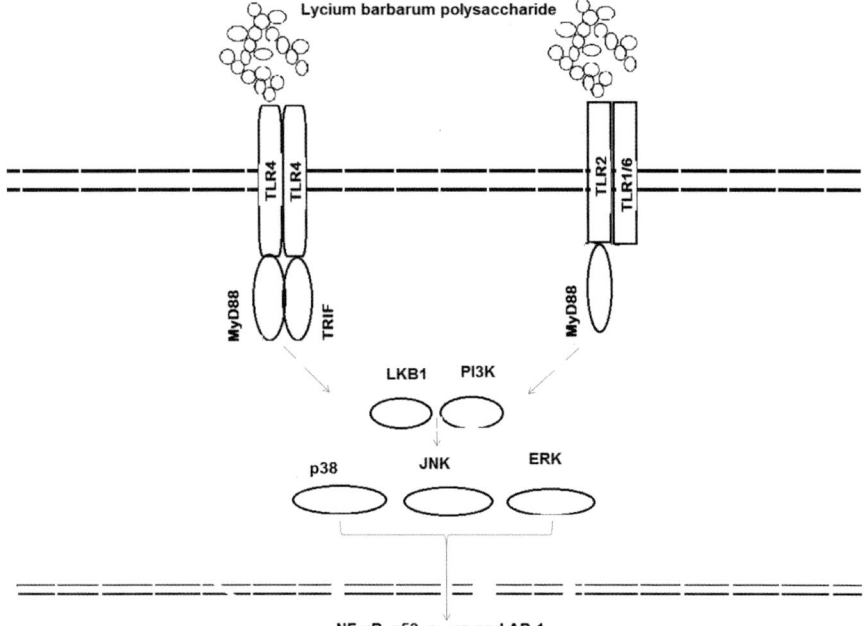

Fig. 2.1 Schematic model illustrating the signaling pathways involved in macrophage activation by *Lycium barbarum* polysaccharides. *L. barbarum* polysaccharides can activate macrophages via Toll-like receptor 4 (*TLR4*) and TLR2. TLR4/2-activated signaling pathways lead to activation of phosphoinositide-3-kinase (*PI3K*) and LKB1, leading to activation of the mitogen-activated protein kinase (*MAPK*), extracellular signal-regulated kinase (*ERK*) and nuclear factor-κB (*NF-κB*), p53, C-Jun, and AP-1. Ultimately, these pathways lead to induction of gene transcription. TLR4 ligation leads to the activation of IL-1R-associated kinase (*IRAK*) via adaptor myeloid differentiation protein 88 (*MyD88*), with subsequent activation of tumor necrosis factor (*TNF*) receptor-associated factor 6 (*TRAF-6*), MAP kinases (*e.g., p38 and JNK*) and NF-κB. Activation of these transcription pathways induces expression of pro-inflammatory cytokines and immune regulation, survival and proliferation

functional maturation of murine BMDC were investigated in vitro. Compared with BMDC in the control group that was exposed to RPMI 1640 only, the co-expression of I-A/I-E, CD11c, and secretion of IL-12 p40 from BMDC were increased by LBPs (100 μg/ml). In addition, the endocytosis of FITC-dextran by LBPs-treated BMDC (100 μg/ml) was impaired, whereas the activation of proliferation of allogenic lymphocytes by BMDC was enhanced. The results strongly suggest that LBPs are capable of promoting both the phenotypic and functional maturation of murine BMDC (Zhu et al. 2007). Both LBPs and polysaccharide-protein complex isolated from *L. barbarum* can induce phenotypic and functional maturation of DCs with strong immunogenicity. Research has demonstrated that LBPs upregulate DC expression of CD40, CD80, CD86, and MHC class II molecules; down-regulate

DC uptake of antigens; enhance DC co-stimulatory activity; and induce IL-12p40 and p70 production. Of all five fractions, LBPF1–5 has been demonstrated to be active. *L. barbarum* polysaccharide-protein complex enhances Th1 responses, and polysaccharide-protein complex-treated DCs enhance Th1 and Th2 responses in vitro and in vivo. The research provides evidence and a rationale for using *L. barbarum* in the treatment of various clinical conditions to enhance host immunity and suggests *L. barbarum* is a potent adjuvant in the design of DC-based vaccines (Chen et al. 2009a).

2.3.1.2 Macrophages

Macrophages play a major role in the host defense against infection. Macrophages express a broad range of pattern recognition receptors (PRRs) to bind the conserved structures of pathogens, ingest bond microbes into vesicles, and produce reactive oxygen intermediates (ROIs) and reactive nitrogen intermediates (mainly nitric oxide) to destroy microbes (Aderem and Underhill 1999). Activated macrophages also secrete the cytokines TNF-α and IL-1 and chemokines to induce inflammatory reactions to microbes (Pylkkanen et al. 2004). In addition, macrophages can present antigen to T cells and produce IL-12 to coordinate innate and adaptive immune responses (Watford et al. 2003). Furthermore, macrophages are involved in tissue remodeling after infections and injury, clearance of apoptotic cells and hematopoiesis (Krysko et al. 2006).

LBPF4-OL is the glycan part of *L. barbarum* polysaccharide-protein complex fraction 4 (LBPF4). A study demonstrated that the LBPF4-OL markedly induced proliferation of spleen cell but could not induce proliferation of purified T and B lymphocytes. Further study revealed that the proliferation of B cell took place in the presence of activated macrophages or LPS. Multiplex bead analysis indicated that LBPF4-OL can obviously induce production of IL-6, IL-8, IL-10 and IFN-α by spleen cells in a concentration-dependent manner. Flow cytometric analysis indicated that LBPF4-OL (i.p.) triggers expression of CD86 and MHC-II on macrophages. An enzyme-linked immunosorbent assay (ELISA) assay demonstrated that LBPF4-OL could greatly stimulate macrophages to secrete TNF-α and IL-1β. These results suggest that the glycan LBPF4-OL plays an important role in the immunopharmacological activity of *L. barbarum* polysaccharide-protein complex; and macrophages, rather than T and B cells, are the principal target cells of LBPF4-OL (Zhang et al. 2011).

It has been found that polysaccharide-protein complex from *L. barbarum* fruit (50 mg/kg, i.p.) markedly upregulated the expressions of CD40, CD80, CD86, and MHC class II molecules on peritoneal macrophages. In vitro studies demonstrated that *L. barbarum* polysaccharide-protein complex activated transcription factors NF-κB and AP-1 in RAW264.7 macrophages; induced mRNA expression for TNF-α, IL-1ß, IL-12p40; and enhanced production of TNF-α in a dose-dependent manner. Furthermore, *L. barbarum* polysaccharide-protein complex (50 mg/kg, i.p.) significantly enhances endocytic and phagocytic capacities of macrophages in an in vivo study. These results indicate that *L. barbarum* polysaccharide-protein complex enhances innate immunity by activating macrophages. The mechanism may be mediated via activation of transcription factors NF-κB and AP-1 to induce production of TNF-α and upregulation of MHC class II (Chen et al. 2009b). The comparisons of immune activities of polysaccharides and polysaccharide-protein complex from *L. barbarum* on macrophages have also been reported. Experiments using in vitro assays found that LBPF4-induced proliferation of splenocytes was dependent on both B and T cells. However, LBPF4-OL-induced splenocyte proliferation was mainly dependent on B cells. The ELISA results indicated that both LBPF4 and LBPF4-OL significantly induced production of TNF-α, IL-1ß, and NO from macrophages. Furthermore, electrophoretic mobility shift assay (EMSA) studies suggest that LBPF4 100 μg/ml can be more effectively to increase NF-kappa B activity than that of LBPF4-OL. The results demonstrate that LBPF4 can enhance T, B cells, and macrophage functions, but LBPF4-OL can only enhance B cells, and macrophage functions. This is partly due to LBPF4 being able to more significantly enhance lymphocytes NF-κB activity (Zhang et al. 2014a).

2.3.1.3   T Lymphocytes

The present research has revealed the effects of *L. barbarum* in activating T cells. Flow cytometry assays revealed that *L. barbarum* polysaccharide enhanced the proliferation of murine splenic lymphocyte. The combined use of LBP and Con A had synergistic effects. MTT assays demonstrated that LBP significantly promoted proliferation of murine splenic lymphocytes, whereas LBP plus Con A combination also enhanced the lymphocyte proliferation at a high concentration. LBP with Con A had effects on immunocompetence (Amagase and Farnsworth 2011). Another research group found that LBP significantly stimulated proliferation of mouse splenocytes. T cell but not B cell proliferation was observed. Cell cycle profile analysis indicated that LBP5 markedly reduced sub-G1 cell expression. LBP could activate the transcription factors NFAT and AP-1, prompt CD25 expression, and induce IL-2 and IFN-γ gene transcription and protein secretion. LBP (i.p. or p.o.) significantly induced proliferation of T cells. The effect of *L. barbarum* glycopeptide 3 (LBGP3) on T cell apoptosis in aged mice has been reported. LBGP3 was purified from *Fructus Lycii* water extracts and identified as a 41 kD glycopeptide. Treatment with 200 μg/ml LBGP3 increased the apoptotic rate of T cells from aged mice and produced a similar DNA ladder pattern to that observed in young T cells. The reversal of apoptotic resistance was involved in down-regulating the expression of Bcl-2 and FLIP and upregulating the expression of FasL. *L. barbarum* glycopeptide 3 reverses apoptotic resistance of aged T cells by modulating the expression of apoptosis-related molecules (Yuan et al. 2008). The results suggest that activation of T lymphocytes by *L. barbarum* may contribute to one of its immune enhancement functions (Chen et al. 2008a).

2.3.1.4 NK Cells

Polysaccharides are believed to be strong immune stimulants that can promote the proliferation and activity of T cells, B cells, macrophages, and NK cells. A study aimed to investigate the effects of polysaccharides, including *L. barbarum* polysaccharide (LBP), on primary human NK cells under normal or simulated microgravity (SMG) conditions. The results demonstrated that LBP markedly promoted the cytotoxicity of NK cells by enhancing the secretion of IFN-γ and perforin and increasing the expression of the activating receptor NKp30 under normal conditions. Moreover, LBP can enhance NK cell function under SMG conditions by restoring the expression of the activating receptor NKG2D and reducing early apoptosis and late apoptosis/necrosis. Additionally, antibody neutralization tests demonstrated that CR3 may be the critical receptor involved in polysaccharide-induced NK cell activation. These findings indicate that polysaccharides may be used as immune regulators to promote the health of the public and even astronauts during space missions (Ting et al. 2014).

2.3.1.5 Other Target Cells

Granular leukocytes and mast cells are the main effector cells of food allergies, which cause type I hypersensitivity. *L. barbarum* berries have been introduced into Western diets. Preliminary reports have demonstrated its allergenic capacity (Ballarin et al. 2011; Larramendi et al. 2012). A study investigated the frequency of sensitization and the allergens. In this study, 566 individuals with respiratory or cutaneous symptoms were skin prick tested with *L. barbarum* berry extract. Thirty-three individuals were positive (5.8 %), and 94 % were sensitized to other allergens. The specific IgE to *L. barbarum* berries, peaches, tomatoes, and a nut mix was measured. Thirteen individual serum samples out of 24 available serum samples (54.2 %) had positive specific IgE. In addition, 92.3 % of *L. barbarum* berry-positive patients were positive to peaches. Seven individuals recognized eight bands, and six recognized a 7-kDa band. This band was identified as a lipid transfer protein by mass spectrometry (MS/MS). Cross-reactivity was demonstrated with tomato, tobacco, nut mix, Artemisia pollen, and purified Lyce3 and Pru p3. These results indicate that *L. barbarum* berries are a new allergenic source with a high prevalence of sensitization (Carnes et al. 2013). Some other reports found that the isolated active component of LBP3a, combined with a DNA vaccine encoding the major outer membrane protein of *Chlamydophila abortus,* induced protection in mice against challenge. A combination of DNA vaccine and LBP3a induced significantly higher antibody levels in mice. MOMP-specific IgG1, IgG2a, and IgG2b antibodies were found in the pool of sera postvaccination on day 42. IgG2a and Ig2b became the predominant isotypes in 12.5, 25, and 50 mg/kg LBP3a-adjuvanted groups (Ling et al. 2011). It has also been reported that mice fed *L. barbarum* had higher influenza antibody titers (Du et al. 2014). These studies indicate that granular leukocytes, mast cells, and B lymphocytes are also related to the *L. barbarum* activity. However, more experiments are required to clarify whether these cells are the direct target cells of *L. barbarum*.

2.3.2 Receptors

Identifying cellular receptors is important to understand how polysaccharides exert their immunomodulatory effects. Several ß-glucan receptors have been identified. The reported ß-glucan receptors include lactosylceramide (LacCer), Toll-like receptors (TLRs) 2 and 6, and dectin-1 (Zimmerman et al. 1998; Sletmoen and Stokke 2008). In addition, TLR4 has been identified as a receptor of polysaccharides, and many polysaccharide activities involve TLR4. Other polysaccharide-related receptors have been reported, including complement receptor-3 (CR3), scavenger receptor (SR), MR (CD206), CD44/RHAMM and selectins.

A study reported that the activity of the polysaccharide LBPF4-OL, which was purified from LBP, is closely associated with the TLR4-MAPK signaling pathway. Research found that LBPF4-OL could significantly induce production of TNF-alpha and IL-1ß in peritoneal macrophages isolated from wild-type (C3H/HeN) but not TLR4-deficient mice (C3H/HeJ). The study also found that the proliferation of LBPF4-OL-stimulated lymphocytes from C3H/HeJ mice is significantly weaker than that of lymphocytes from C3H/HeN mice. Furthermore, through a bio-layer interferometry assay, it was found that LPS but not LBPF4-OL can directly associate with the TLR4/MD2 molecular complex. Flow cytometry analysis indicated that LBPF4-OL markedly upregulates TLR4/MD2 expression in both peritoneal macrophages and Raw264.7 cells. As its mechanism of action, LBPF4-OL increases the phosphorylation of p38-MAPK and inhibits the phosphorylation of JNK and ERK1/2, as was examined by western blot analysis. These data suggest that the *L. barbarum* polysaccharide LBPF4-OL is a new Toll-like receptor 4/MD2-MAPK signaling pathway activator and inducer (Zhang et al. 2014b). Similar results have been observed for dendritic cells. Zhu et al. reported that LBPs induced phenotypic and functional maturation of DCs. LBPs upregulated DC expression of I-A/I-E and CD11c, enhanced DC allostimulatory activity and induced production of IL-12p40. Furthermore, the activity of LBPs on DCs was significantly reduced by treating the cells with anti-TLR2 or anti-TLR4 antibody prior to LBPs, indicating that both are possible receptors of LBPs. Maturation of DCs by LBPs was able to directly activate the nuclear transcription factor NF-κB p65. The results revealed that LBP stimulation induces the phenotypic and functional maturation of DCs via TLR2 and/or TLR4-mediated NF-κB signaling pathways (Zhu et al. 2013). The above results indicate the immunoactivity of *L. barbarum* is related to TLR4/2. Whether other receptors are related to *L. barbarum* remains unknown.

2.3.3 Signal Transduction

Several signal transduction pathways, including the PI3K/Akt/FoxO1, LKB1/ AMPK, JNK/c-Jun, MEK/ERK, and PI3K/HIF-1α pathways, and transcription factors NF-κB, p53, c-Jun, and AP-1 are reported to be *L. barbarum* activity-related signal transduction molecules.

L. barbarum polysaccharides (LBPs) from wolfberries have been reported to have antioxidant and neuroprotective derivatives. A study found that LBPs are also a novel hepatoprotective agent against nonalcoholic steatohepatitis (NASH) caused by a diet-induced NASH rat model. The study examined female rats fed with 1 mg/ kg LBP daily for 8 weeks and compared with control rats. NASH+ LBPs-cotreated rats displayed (1) improved histology and free fatty acid levels, (2) re-balancing of lipid metabolism, (3) reducing profibrogenic factors through the TGF-ß/SMAD pathway, (4) improved oxidative stress through the cytochrome P450 2E1-dependent pathway, (5) reducing production of hepatic pro-inflammatory mediators and chemokines, and (6) ameliorating hepatic apoptosis through the p53-dependent intrinsic and extrinsic pathways. All these effects of LBP were partly modulated through the PI3K/Akt/FoxO1, LKB1/AMPK, JNK/c-Jun, and MEK/ERK pathways and the down-regulation of transcription factors in the liver, such as NF-κB and activator protein-1 (AP-1) (Xiao et al. 2013). Moreover, LBPs have also been found to inhibit tumor cell growth by suppressing IGF-1-induced angiogenesis via PI3K/ HIF-1α/VEGF signaling pathways. Studies have reported that a 90 h treatment with 0.50 mg/ml of LBPs resulted in significant inhibition of MCF-7 cell proliferation. Using this same cell type, studies have also observed that LBPs could also affect insulin-like growth factor (IGF)-1 protein accumulation, suppress PI3K activity and p-PI3K protein levels, inhibit accumulation of hypoxia-inducible factor-1 (HIF-1α) protein without altering HIF-1α mRNA levels, and suppress mRNA expression and protein production of VEGF (Huang et al. 2012).

NF-κB has also been found to be one of the most important transcription factors related to *L. barbarum* activity. A study reported that LBP treatment may protect against intestinal ischemia-reperfusion injury (IRI)-induced intestinal damage by inhibiting PMN accumulation and ICAM-1 expression and ameliorating changes in TNF-α level, NF-κB activation, intestinal permeability, and histology (Yang et al. 2013). Other reports indicate that LBPs do not delay primary degeneration of RGCs after either complete optic nerve transection (CONT) or partial optic nerve transection (PONT), but they delay secondary degeneration of retinal ganglion cells (RGCs) after PONT. The study found that LBPs appeared to exert these protective effects by inhibiting oxidative stress and the JNK/c-Jun pathway and by transiently increasing production of insulin-like growth factor-1 (IGF-1) (Li et al. 2013). After investigating the effect of LBP on the differentiation and maturation of healthy human peripheral blood-derived dendritic cells cultured in different tumor microenvironments in vitro and evaluating the molecular and immunological mechanisms of LBP in the treatment of tumors, a study reported that LBPs could increase the expression of the phenotype of DCs, secretion of IL-12p70 and IFN-γ in MLR and enhance NF-κB expression, especially in virus-related peripheral blood-derived dendritic cell precursor cells, suggesting that LBPs play a stronger antitumor role in virus-related environments, and this phenomenon correlates with the NF-κB signaling pathway (Chen et al. 2012). Maturation of DCs by LBPs is able to directly activate the nuclear transcription factor NF-κB p65 (Zhu et al. 2013).

2.3.4 Cytokines

As a type of immune regulator, *L. barbarum* can induce production of a variety of cytokines, such as the anti-inflammatory factor IL-10, proinflammatory cytokines IL-1ß and IL-6, chemokines IL-8, antitumor factors cytokine TNF-α, antiviral factor IFN-γ, TGF-ß1, and lymphocyte activators IL-2 and IL-4. LBPF4-OL is the glycan part of *L. barbarum L* polysaccharide-protein complex fraction 4 (LBPF4). A study found that LBPF4-OL can obviously induce production of IL-6, IL-8, IL-10, and INF-α from mouse spleen cells in a concentration-dependent manner in vitro (Zhang et al. 2011). The effects of *L. barbarum* polysaccharide-protein complex (LBP3p) on the expression of IL-2 and TNF-α in human peripheral blood mononuclear cells have been examined by reverse transcription polymerase chain reaction (RT-PCR) and bioassay. A study found that administration of LBP3p increased the expression of IL-2 and TNF-α at both the mRNA and protein levels in a concentration-dependent manner (Gan et al. 2003). LBPs can also increase the secretion of IL-12 p40 of BMDC in vitro (Zhu et al. 2007). LBP, LBPF4, and LBPF5 have been reported to significantly stimulate proliferation of mouse splenocyte and induce gene expression for IL-2 and IFN-γ as well as their protein secretion (Chen et al. 2008a). A study also found adoptive transfer of wolfberry-treated bone marrow DCs (loaded with ovalbumin (323–339)-peptide) promoted proliferation of antigen-specific T cells as well as production of interleukin-4 and interferon-gamma in CD4(+) T cells (Du et al. 2014). The above results indicate LBPs can induce cytokine production in vitro.

It has been reported that polysaccharide-rich *L. barbarum* and *Rehmannia glutinosa* treatment increases hepatic anti-inflammatory cytokine IL-10 levels, suppresses liver fibrosis-biomarkers TGF-ß1, and reduces hepatic levels of the pro-inflammatory cytokines TNF-α and IL-1ß after exposing the rats to carbon tetrachloride (CCl_4) (Wu et al. 2011). Another study fed adult mice (4 month-old) a milk-based preparation of wolfberries called Lacto-Wolfberry (LWB) for 4 weeks and then infected the mice with influenza A/Puerto Rico/8/34 (HI NI) while continuing the same experimental diet. The LWB-fed mice displayed, overall, significantly higher concanavalin A-induced IL-2 production. Furthermore, the study found positive correlations between weight loss, lung viral titer, pathology score, TNF-α, and IL-6 production, as well as negative correlations with T cell proliferation and IL-2 production (Ren et al. 2012). The aforementioned results indicate LBPs can induce cytokine production in vivo.

2.3.5 Antibodies

L. barbarum polysaccharides (LBPs) can moderate immune responses. They could potentially be used as a substitute for oil adjuvants in veterinary vaccines. It has been demonstrated that the isolated active component of LBP3a, combined with a DNA vaccine encoding the major outer membrane protein of *C. abortus*, exerted protection to mice against the challenge. A combination of DNA vaccine and LBP3a significantly induced higher levels of antibodies in mice. MOMP-specific IgG1, IgG2a, and IgG2b antibodies were found in a sera pool 42 days after vaccination. IgG2a and Ig2b became the predominant isotypes in 12.5, 25, and 50 mg/kg LBP3a-adjuvanted groups (Ling et al. 2011). Sulfated *L. barbarum* polysaccharides (sLB-PSs) with different degrees of sulfation (DS), sLBPS (1.5), and sLBPS (1.9) were added into cultured chicken peripheral lymphocytes, and the change of lymphocyte proliferation was compared by MTT assay, taking the nonmodified LBPS as a control. On days 7, 14, 21, and 28, after the first vaccination, the changes in proliferation of peripheral lymphocyte and serum hemagglutination inhibition (HI) antibody titer were determined. The results indicated that two sLBPSs could significantly promote proliferation of lymphocyte and enhance serum antibody titer (Wang et al. 2010b). A 3-month randomized, double-blinded, placebo-controlled study was conducted on 150 healthy community-dwelling Chinese elderly (65–70 years old) supplemented with Lacto-Wolfberry or placebo (13.7 g/day). The immune response to influenza vaccine was assessed in the study, along with inflammatory and physical status. No serious adverse reaction was reported during the trial nor were symptoms of influenza-like infection. No changes in body weight and blood pressure, blood chemistry or cell composition, as well as in autoantibodies levels were observed. The subjects receiving Lacto-Wolfberry had significantly higher levels of serum influenza-specific immunoglobulin G after vaccination and the rate of seroconversion between days 30 and 90 as compared with the placebo group. The postvaccination positive rate was greater in the Lacto-Wolfberry group than the placebo group but did not reach significance (Vidal et al. 2012). It has also been reported that mice fed with wolfberries had higher influenza antibody titers (Du et al. 2014). The effects of *L. barbarum* polysaccharide (LBP) on immune responses in vaccinated chickens were also reported. A total of 600 Hy-Line Brown chickens aged 15 days old were randomly divided into four groups with three replicates per group and 50 chickens per replicate, and all of the chickens were injected with Newcastle Disease (ND) vaccine. Three experimental groups of chickens were injected with 20, 10, and 5 mg/kg LBP (LBPH, LBPM, and LBPL). The results indicated that LBP (10 and 20 mg/kg) could significantly enhance the ND antibody titers (Qin et al. 2012).

2.3.6 Leukocyte Differentiation Antigen

DC pheonotypic maturation is related to its immunogenicity. The priming of T cells requires that both peptide-MHC complex and CD80/CD86 on APCs bind to TCR and CD28 on T cells, respectively. To determine whether LBP induces DC maturation, a study generated DCs from BALB/c mouse bone marrow and stimulated them with LBPs or individual fractions. Twenty-four hour treatment with LBPs or LBPF1–5 upregulated CD40, CD80, CD86, and MHC class II molecules to various extents compared with the medium control. The increase in mean fluorescence intensity was most prominent for MHC-II and CD86 by LBP, BLPF4, or LBPF5 stimulations. CD40 expression was weak, but there was still a ~2-fold increase after treatment. Surprisingly, LBP (s.c.) was most effective, perhaps because the in situ activated DCs migrate to the spleen more effectively through the lymphatic system (Chen et al. 2009a). Compared with the BMDC that were only subjected to treatment with RPM11640, the coexpression of I-A/I-E and CD11c by bone marrow-derived dendritic cells stimulated with LBPs (100 µg/ml) were increased (Zhu et al. 2007).

2.4 Future Remarks and Conclusions

Thus far, a profound understanding of the immune regulating function of *L. barbarum* has been achieved, especially the antitumor and immune regulation mechanism of LBPs, includes deep knowledge of the receptors and signal transduction levels. Studies have found that the effect of LBPs on the activation of macrophages and dendritic cells are important in participating in the immune response. In addition, it was found that TLR4/2 may be closely related to the immunoregulatory activity of LBPs. In addition, whether other PRRs, which include SRs, mannose receptors, TLRs, Dectin-1, and complement receptor type 3 (CR3), are associated with the activity of *L. barbarum* is not yet known. It will be interesting to identify the receptor of *L. barbarum* on macrophages or dendritic cells in the future.

Little is known about the LBP protective effects on neurons, the anti-neurogenesis activity, and the effect on the drug-induced learning and memory mechanism. A recent study found that LBPs can prevent scopolamine-induced cognitive and memory deficits and reductions in cell proliferation and neuroblast differentiation (Chen et al. 2014). Other research indicates that LBPs can delay secondary degeneration of RGCs, and this effect may be linked to inhibition of oxidative stress and the JNK/c-Jun pathway in the retina and transient increases in production of insulin-like growth factor-1 (IGF-1) (Li et al. 2013). In addition, there is also a research report indicated that LBP treatment may protect against IRI-induced intestinal damage, possibly by inhibiting IRI-induced oxidative stress and inflammation (Yang et al. 2013). Recently, a new study reported that LBPs partially exerted their beneficial neuroprotective effects on ischemia-reperfusion via the activation of Nrf2 and an increase in HO-1 protein expression (He et al. 2014). These studies suggest that antioxidant effects may play a key role in the LBP protective effect on various diseases. Exploration of the effect of *L. barbarum* on the influence of the balance of the body's REDOX reactions, such as key oxidative stress Nrf-2/ARE pathways, may further reveal the intrinsic relationship *of L. barbarum*'s antioxidant and immune regulation functions.

References

Aderem A, Underhill DM. Mechanisms of phagocytosis in macrophages. Ann Rev Immunol. 1999;17:593–623. doi:10.1146/annurev.immunol.17.1.593.

Altintas A, Kosar M, Kirimer N, Baser KHC, Demirci B. Composition of the essential oils of *Lycium barbarum* and L-ruthenicum fruits. Chem Nat Compd. 2006;42(1):24–5. doi:10.1007/s10600-006-0028-3.

Amagase H, Farnsworth NR. A review of botanical characteristics, phytochemistry, clinical relevance in efficacy and safety of *Lycium barbarum* fruit (Goji). Food Res Int. 2011;44(7):1702–17. doi:10.1016/j.foodres.2011.03.027.

Amagase H, Sun B, Nance D.:Immune modulation by a standardized *Lycium barbarum* fruit (Goji) juice in randomized, double-blind, placebo-controlled clinical studies.Paper presented at the 7th Joint Meeting of the Association-Francophone-pour-l'Enselgnement-et-la-Recherche-en Pharmacognosie/American-Society-of-Pharmacognosy/Society-for-Medicinal-Plant-Research/Phytochem-Society-of-Europe/Societa-Italiana-di-Fitochimica, Athens, Greece, 3–8 Aug, 2008.

Archer GS, Mench JA. The effects of the duration and onset of light stimulation during incubation on the behavior, plasma melatonin levels, and productivity of broiler chickens. J Anim Sci. 2014;92(4):1753–8. doi:10.2527/jas2013-7129.

Ballarin SM, Lopez-Matas MA, Abad DS, Perez-Cinto N, Carnes J. Anaphylaxis associated with the ingestion of goji berries (*Lycium barbarum*). J Invest Allergol Clin Immunol. 2011;21(7):567–70.

Banchereau J, Briere F, Caux C, Davoust J, Lebecque S, Liu YT, Pulendran B, Palucka K. Immunobiology of dendritic cells. Annu Rev Immunol. 2000;18:767–811. doi:10.1146/annurev.immunol.18.1.767.

Bhalla Y, Gupta VK, Jaitak V. Anticancer activity of essential oils: a review. J Sci Food Agric. 2013;93(15):3643–53. doi:10.1002/jsfa.6267.

Carnes J, de Larramendi CH, Ferrer A, Huertas AJ, Lopez-Matas MA, Pagan JA, Navarro LA, Garcia-Abujeta JL, Vicario S, Pena M. Recently introduced foods as new allergenic sources: sensitisation to goji berries (*Lycium barbarum*). Food Chem. 2013;137(1–4):130–5. doi:10.1016/j.foodchem.2012.10.005.

Chan HC, Chang RCC, Ip AKC, Chiu K, Yuen WH, Zee SY, So KF. Neuroprotective effects of *Lycium barbarum* Lynn on protecting retinal ganglion cells in an ocular hypertension model of glaucoma. Exp Neurol. 2007;203(1):269–73. doi:10.1016/j.expneurol.2006.05.031.

Chen Z, Kwong Huat Tan B, Chan SH. Activation of T lymphocytes by polysaccharide-protein complex from *Lycium barbarum* L. Int immunopharmacol. 2008a;8(12):1663–71. doi:10.1016/j.intimp.2008.07.019.

Chen ZS, Lu JH, Srinivasan N, Tan BKH, Chan SH. Polysaccharide-protein complex from *Lycium barbarum* L. Is a novel stimulus of dendritic cell immunogenicity. J Immunol. 2009a;182(6):3503–9. doi:10.4049/jimmunol.0802567.

Chen ZS, Soo MY, Srinivasan N, Tan BKH, Chan SH. Activation of macrophages by polysaccharide-protein complex from *Lycium barbarum* L. Phytother Res. 2009b;23(8):1116–22. doi:10.1002/ptr.2757.

Chen JR, Li EQ, Dai CQ, Yu B, Wu XL, Huang CR, Chen XY. The inducible effect of LBP on maturation of dendritic cells and the related immune signaling pathways in Hepatocellular Carcinoma (HCC). Curr Drug Deliv. 2012;9(4):414–20.

Chen WW, Cheng X, Chen JZ, Yi X, Nie DK, Sun XH, Qin JB, Tian ML, Jin GH, Zhang XH. *Lycium barbarum* polysaccharides prevent memory and neurogenesis impairments in scopol-amine-treated rats. PloS ONE. 2014;9(2):e88076. doi:10.1371/journal.pone.0088076.

Cuevas A, Saavedra N, Salazar LA, Abdalla DSP. Modulation of immune function by polyphe-nols: possible contribution of epigenetic factors. Nutrients. 2013;5(7):2314–32. doi:10.3390/nu5072314.

Dong JZ, Gao WS, Lu DY, Wang Y. Simultaneous extraction and analysis of four polyphenols from leaves of *Lycium barbarum* L. J Food Biochem. 2011;35(3):914–31. doi:10.1111/j.1745-4514.2010.00429.x

Dong JZ, Wang Y, Wang SH, Yin LP, Xu GJ, Zheng C, Lei C, Zhang MZ. Selenium increases chlorogenic acid, chlorophyll and carotenoids of Lycium chinense leaves. J Sci Food Agric. 2013;93(2):310–5. doi:10.1002/jsfa.5758.

Du XG, Wang JP, Niu XL, Smith D, Wu DY, Meydani SN. Dietary wolfberry supplementation enhances the protective effect of flu vaccine against influenza challenge in aged mice. J Nutr. 2014;144(2):224–9. doi:10.3945/jn.113.183566.

Gan L, Zhang SH, Liu Q, Xu HB. A polysaccharide-protein complex from *Lycium barbarum* up-regulates cytokine expression in human peripheral blood mononuclear cells. Eur J Pharmacol. 2003;471(3):217–22. doi:10.1016/s0014-2999(03)01827-2.

Gan L, Hua Zhang S, Liang Yang X, Bi Xu H. Immunomodulation and antitumor activity by a poly-saccharide-protein complex from *Lycium barbarum*. Int immunopharmacol. 2004;4(4):563–9. doi:10.1016/j.intimp.2004.01.023.

Gao ZP, Ali Z, Khan IA. Glycerogalactolipids from the fruit of *Lycium barbarum*. Phytochemistry. 2008;69(16):2856–61. doi:10.1016/j.phytochem.2008.09.002.

Guermonprez P, Valladeau J, Zitvogel L, Thery C, Amigorena S. Antigen presentation and T cell stimulation by dendritic cells. Annu Rev Immunol. 2002;20:621–67. doi:10.1146/annurev.im-munol.20.100301.064828.

He MH, Pan H, Chang RCC, So KF, Brecha NC, Pu ML. Activation of the Nrf2/HO-1 antioxidant pathway contributes to the protective effects of *Lycium barbarum* polysaccharides in the rodent retina after ischemia-reperfusion-induced damage. PloS ONE. 2014;9(1):e84800. doi:10.1371/journal.pone.0084800.

Ho YS, So KF, Chang RCC. Anti-aging herbal medicine-how and why can they be used in aging-associated neurodegenerative diseases? Ageing Res Rev. 2010;9(3):354–62. doi:10.1016/j.arr.2009.10.001.

Huang LJ, Tian GY, Qi CH, Zhang YX. Structure elucidation and immunoactivity studies of gly-can of glycoconjugate LbGp4 isolated from the fruit of *Lycium barbarum* L. Chem J Chin Univ-Chin. 2001;22(3):407–11.

Huang X, Zhang QY, Jiang QY, Kang XM, Zhao L. Polysaccharides derived from *Lycium bar-barum* suppress IGF-1-induced angiogenesis via PI3K/HIF-1 alpha/VEGF signalling pathways in MCF-7 cells. Food Chem. 2012;131(4):1479–84. doi:10.1016/j.foodchem.2011.10.039.

Inbaraj BS, Lu H, Kao TH, Chen BH. Simultaneous determination of phenolic acids and fla-vonoids in *Lycium barbarum* Linnaeus by HPLC-DAD-ESI-MS. J Pharm Biomed Anal. 2010;51(3):549–56. doi:10.1016/j.jpba.2009.09.006.

Ip AKC, Chiu K, Yu MS, Yuen WH, Zee SY, Chang RCC, So KF. Neuroprotective effect of *Lycium barbarum* in rat chronic ocular hypertension model via immunomodulation of macrophages/microglia. Neurosignals. 2006;15(3):145–6.

Krysko DV, D'Herde K, Vandenabeele P. Clearance of apoptotic and necrotic cells and its immu-nological consequences. Apoptosis. 2006;11(10):1709–26. doi:10.1007/s10495-006-9527-8.

Larramendi CH, Garcia-Abujeta JL, Vicario S, Garcia-Endrino A, Lopez-Matas MA, Garcia-Sede-no MD, Carnes J. Goji berries (*Lycium barbarum*): risk of allergic reactions in individuals with food allergy. J Invest Allergol Clin Immunol. 2012;22(5):345–50.

Li XM, Ma YL, Liu XJ. Effect of the *Lycium barbarum* polysaccharides on age-related oxidative stress in aged mice. J Ethnopharmacol. 2007;111(3):504–11. doi:10.1016/j.jep.2006.12.024.

Li HY, Liang YX, Chiu K, Yuan QJ, Lin B, Chang RCC, So KF. *Lycium barbarum* (wolfberry) reduces secondary degeneration and oxidative stress, and inhibits JNK pathway in retina after partial optic nerve transection. PloS ONE. 8(7):e0068881. doi:10.1371/journal.pone.0068881.

Ling Y, Li SW, Yang JJ, Yuan JL, He C. Co-administration of the polysaccharide of *Lycium barbarum* with DNA vaccine of Chlamydophila abortus augments protection. Immunol Invest. 2011;40(1):1–13. doi:10.3109/08820139.2010.504803.

Liu HH, Fan YL, Wang WH, Liu NA, Zhang H, Zhu ZY, Liu AJ. Polysaccharides from *Lycium barbarum* leaves: isolation, characterization and splenocyte proliferation activity. Int J Biol Macromol. 2012;51(4):417–22. doi:10.1016/j.ijbiomac.2012.05.025.

Nan Y, Wang R, Yuan L, Niu Y. Effects of *Lycium barbarum* polysaccharides (LBP) on immune function of mice. Afr J Microbiol Res. 2012;6(22):4757–60. doi:10.5897/ajmr12.363.

Park SY, Park WT, Park YC, Ju JI, Park SU, Kim JK. Metabolomics for the quality assessment of Lycium chinense fruits. Biosci Biotechnol Biochem. 2012;76(12):2188–94. doi:10.1271/bbb.120453.

Peng XM, Qi CH, Tian GY, Zhang YX. Physico-chemical properties and bioactivities of a glycoconjugate LbGp5B from *Lycium barbarum* L. Chin J Chem. 2001;19(9):842–6.

Pylkkanen L, Gullsten H, Majuri ML, Andersson U, Vanhala E, Maatta J, Meklin T, Hirvonen MR, Alenius H, Savolainen K. Exposure to Aspergillus fumigatus spores induces chemokine expression in mouse macrophages. Toxicology. 2004;200(2–3):255–63. doi:10.1016/j.tox.2004.03.019.

Qin RL, Chu YC, Lian JF, Xu ZY, Li CH, Wang L, Li GH. Immunoregulation of *Lycium barbarum* polysaccharide in vaccinated chickens. J Anim Vet Adv. 2012;11(17):3105–10.

Ren ZH, Na LX, Xu YM, Rozati M, Wang JP, Xu JG, Sun CH, Vidal K, Wu DY, Meydani SN. Dietary supplementation with lacto-wolfberry enhances the immune response and reduces pathogenesis to influenza infection in mice. J Nutr. 2012;142(8):1596–602. doi:10.3945/jn.112.159467.

Sletmoen M, Stokke BT. Review: higher order structure of (1,3)-beta-D-glucans and its influence on their biological activities and complexation abilities. Biopolymers. 2008;89(4):310–21. doi:10.1002/bip.20920.

Tian GY, Wang C, Feng YC. Isolation, purification and properties of lbgp and characterization of its glycan-peptide bond. Acta Biochim Biophsy Sin. 1995;27(2):201–6.

Ting HY, Li Q, Yang H, Jin ML, Zhang MJ, Ye LJ, Li J, Huang QS, Yin DC. Protective effect of polysaccharides on simulated microgravity-induced functional inhibition of human NK cells. Carbohydr Polym. 2014;101:819–27. doi:10.1016/j.carbpol.2013.10.021.

Vidal K, Bucheli P, Gao QT, Moulin J, Shen LS, Wang JK, Blum S, Benyacoub J. Immunomodulatory effects of dietary supplementation with a milk-based wolfberry formulation in healthy elderly: a randomized, double-blind, placebo-controlled trial. Rejuv Res. 2012;15(1):89–97. doi:10.1089/rej.2011.1241.

Wang CC, Chang SC, Inbaraj BS, Chen BH. Isolation of carotenoids, flavonoids and polysaccharides from *Lycium barbarum* L. and evaluation of antioxidant activity. Food Chem. 2010a;120(1):184–92. doi:10.1016/j.foodchem.2009.10.005.

Wang JM, Hu YL, Wang DY, Liu J, Zhang J, Abula S, Zhao BA, Ruan SL. Sulfated modification can enhance the immune-enhancing activity of *lycium barbarum* polysaccharides. Cell Immunol. 2010b;263(2):219–23. doi:10.1016/j.cellimm.2010.04.001.

Wang K, Sasaki T, Li W, Li Q, Wang YH, Asada Y, Kato H, Koike K. Two novel steroidal alkaloid glycosides from the seeds of *Lycium barbarum*. Chem Biodivers. 2011;8(12):2277–84. doi:10.1002/cbdv.201000293.

Watford WT, Moriguchi M, Morinobu A, O'Shea JJ. The biology of IL-12: coordinating innate and adaptive immune responses. Cytokine Growth Factor Rev. 2003;14 (5):361–8. doi:10.1016/s1359-6101(03)00043-1.

Web of Science, US Pike, Rockville. http://apps.webofknowledge.com/full_record.do?product=WOS&search_mode=GeneralSearch&qid=4&SID=N2xWbEc76nZW4JsggYJ&page=1&doc=1. Accessed 22 Apr 2008.

Wilson NS, Villadangos JA. Regulation of antigen presentation and cross-presentation in the dendritic cell network: facts, hypothesis, and immunological implications. In: Alt FW, editor. Advances in immunology, Vol. 86. Amsterdam: Elsevier; 2005. pp. 241–305. doi:10.1016/s0065-2776(04)86007-3.

Winzler C, Rovere P, Rescigno M, Granucci F, Penna G, Adorini L, Zimmermann VS, Davoust J, RicciardiCastagnoli P. Maturation stages of mouse dendritic cells in growth factor-dependent long-term cultures. J Exp Med. 1997;185(2):317–28. doi:10.1084/jem.185.2.317.

Wu PS, Wu SJ, Tsai YH, Lin YH, Chao JCJ. Hot water extracted *Lycium barbarum* and Rehmannia glutinosa inhibit liver inflammation and fibrosis in rats. Am J Chin Med. 2011;39(6):1173–91. doi:10.1142/s0192415x11009482.

Wu WB, Hung DK, Chang FW, Ong ET, Chen BH. Anti-inflammatory and anti-angiogenic effects of flavonoids isolated from *Lycium barbarum* Linnaeus on human umbilical vein endothelial cells. Food Funct. 2012;3(10):1068–81. doi:10.1039/c2fo30051f.

Xiao J, Liong EC, Ching YP, Chang RCC, Fung ML, Xu AM, So KF, Tipoe GL. *Lycium barbarum* polysaccharides protect rat liver from non-alcoholic steatohepatitis-induced injury. Nutr Diabetes. 2013;3:e81. doi:10.1038/nutd.2013.22.

Yang XK, Bai H, Cai WX, Li J, Zhou Q, Wang YC, Han JT, Zhu XX, Dong ML, Hu DH. *Lycium barbarum* polysaccharides reduce intestinal ischemia/reperfusion injuries in rats. Chem-Biol Interact. 2013;204(3):166–72. doi:10.1016/j.cbi.2013.05.010.

Yuan LG, Deng HB, Chen LH, Li DD, He QY. Reversal of apoptotic resistance by *Lycium barbarum* glycopeptide 3 in aged T cells. Biomed Environ Sci. 2008;21(3):212–7. doi:10.1016/s0895-3988(08)60031-8.

Zhang XR, Zhou WX, Zhang YX, Qi CH, Yan HG, Wang ZF, Wang B. Macrophages, rather than T and B cells are principal immunostimulatory target cells of *Lycium barbarum* L. polysaccharide LBPF4-OL. J Ethnopharmacol. 2011;136(3):465–72. doi:10.1016/j.jep.2011.04.054.

Zhang X, Li Y, Cheng J, Liu G, Qi C, Zhou W, Zhang Y. Immune activities comparison of polysaccharide and polysaccharide-protein complex from *Lycium barbarum* L. Int J Biol Macromol. 2014a;65C:441–5. doi:10.1016/j.ijbiomac.2014.01.020.

Zhang XR, Qi CH, Cheng JP, Liu G, Huang LJ, Wang ZF, Zhou WX, Zhang YX. *Lycium barbarum* polysaccharide LBPF4-OL may be a new toll-like receptor 4/MD2-MAPK signaling pathway activator and inducer. Int immunopharmacol. 2014b;19(1):132–41. doi:10.1016/j.intimp.2014.01.010.

Zhao CJ, He YQ, Li RZ, Cui GH. Chemistry and pharmacological activity of peptidoglycan from *Lycium barbarum* l. Chin Chem Lett. 1996;7(11):1009–10.

Zhu J, Zhao LH, Zhao XP, Chen Z. Lycium barbarum polysaccharides regulate phenotypic and functional maturation of murine dendritic cells. Cell Biol Int. 2007;31(6):615–9. doi:10.1016/j.cellbi.2006.12.002.

Zhu J, Zhang YY, Shen YS, Zhou HQ, Yu XM. *Lycium barbarum* polysaccharides induce toll-like receptor 2-and 4-mediated phenotypic and functional maturation of murine dendritic cells via activation of NF-kappa B. Mol Med Rep. 2013;8(4):1216–20. doi:10.3892/mmr.2013.1608.

Zimmerman JW, Lindermuth J, Fish PA, Palace GP, Stevenson TT, DeMong DE. A novel carbohydrate-glycosphingolipid interaction between a beta-(1-3)-glucan immunomodulator, PGG-glucan, and lactosylceramide of human leukocytes. J Biol Chem. 1998;273(34):22014–20. doi:10.1074/jbc.273.34.22014.

Chapter 3
The Antioxidant, Anti-inflammatory, and Antiapoptotic Effects of Wolfberry in Fatty Liver Disease

Jia Xiao and George L. Tipoe

Abstract The liver is one of the most important solid organs in the human body, which is responsible for a spectrum of key physiological processes, including protein synthesis, detoxification, nutrient storage, and innate/adaptive immunity regulation. Acute and chronic liver diseases greatly threaten the healthy condition of the liver organ. Alternative medicinal therapy, using extracts from natural herbs, provides a promising way to treat and prevent various liver diseases in past decades. Wolfberry has been proposed as a liver-nourishing recipe in the traditional Chinese medicine for more than 1000 years. Modern preclinical studies and clinical trials also suggest their beneficial effects on the liver. The hepatic protective properties of wolfberry may be mainly attributed to its polysaccharides portion, which have demonstrated to ameliorate hepatic injuries caused by both acute and chronic liver diseases, through mechanisms at different cascading levels of cell signaling pathways. In this chapter, we mainly focus on the etiology of fatty liver diseases and the protective mechanisms of wolfberry on the pathogenesis of chronic liver diseases.

Keywords Fatty liver disease · Wolfberry · Oxidative stress · Inflammation · Apoptosis

3.1 Introduction

The liver is the largest internal organ and gland in the human body. Anatomically, it is mainly situated at the upper right quadrant of the abdominal cavity although a portion of it may extend to the left abdominal cavity. Human normal liver weighs 1.44–1.66 kg, heavier in men as compared to women (Abdel-Misih and Bloomston 2010). The liver is connected to two large vessels—one is hepatic artery which

G. L. Tipoe (✉)
Department of Anatomy, The University of Hong Kong, Hong Kong SAR,
People's Republic of China
e-mail: tgeorge@hku.hk

J. Xiao
Department of Immunobiology, Institute of Tissue Transplantation and Immunology,
Jinan University, Guangzhou, People's Republic of China

© Springer Science+Business Media Dordrecht 2015   45
R. C-C. Chang, K-F. So (eds.), *Lycium Barbarum and Human Health*,
DOI 10.1007/978-94-017-9658-3_3

carries blood from the aorta and the other is portal vein which drains blood from the gastrointestinal tract and spleen to the hepatic capillaries. There are a large number of constant and essential functions served by the liver to maintain homeostasis of the body. The main functions of this organ include protein synthesis, detoxification, storage of nutrition, production of urea, and breakdown of fats (Ishibashi et al. 2009). Therefore, malfunctions of the liver caused by either acute or chronic injury may lead to fatal outcomes.

Common liver diseases can be divided into two categories: acute and chronic. Acute liver injury is the appearance of liver cell damaged that happens rapidly in a short period of time. Severe acute liver injury, if not successfully controlled, may progress to acute liver failure which is a rare clinical condition. According to the onset time of jaundice to encephalopathy, acute liver injury (or failure) can be divided into hyperacute (within 7 days), acute (between 8 and 28 days), and subacute (between 4 and 12 weeks) (O'Grady et al. 1993). Acute liver failure is associated with a high mortality rate up to 80 % (Gill and Sterling 2001). The etiology of acute liver injury/failure could be very complicated. Common causes include drug hepatotoxicity, viral infection, autoimmune-related hepatitis, sepsis, and Wilson's disease. In the USA, the most common cause of acute injury/failure is acetaminophen toxicity, followed by other drug-related injuries (Larson et al. 2005). In addition, in some cases, it is not easy to determine the cause of acute liver injury/failure. For example, in a report of 308 acute liver failure cases, 53 cases are indeterminate. Moreover, those patients with undetermined cause of acute liver failure tend to have lower survival rates than those with more prevalent causes (Ostapowicz et al. 2002). The inability to determine the causes of these diseases may be due to insufficient history taking, unknown viral infection, and the availability of diagnosis (Lee 2008).

Rapid recognition and diagnosis of acute liver injury/failure is critical for the treatment of the disease. The clinical symptoms of acute liver injury/failure range from mild gastrointestinal upset to malaise or even confusion (Polson and Lee 2005). Physical examination and detailed patient history should be assessed firstly to identify the underlying cause. Chronic liver disease may provide clues for the acute disease as well (Czaja and Freese 2002; Horton et al. 2008; Korman et al. 2008). Mental status of patients should be carefully monitored since the change of neurological status may induce coma. In the early course of acute liver injury, jaundice may not be very obvious because patients may still preserve adequate amount of liver function. However, in the later course, serum bilirubin level would be elevated together with alanine transaminase (ALT) and aspartate transaminase (AST) levels due to impaired hepatic metabolism. Other serological and laboratory tests, such as complete blood count, basic metabolic panel, acute hepatitis panel, ammonia test, human immunodeficiency virus (HIV) test, and urine toxicity, are helpful in determining the exact cause of acute liver injury. When laboratory tests are inconclusive, liver biopsy may provide a solid basis for diagnosis (Larson 2010; Miraglia et al. 2006).

It is very critical to understand the mechanisms between the liver cell damage and regeneration when acute liver injury happens. The main consequence of acute liver injury is cellular death induced by oxidative stress or inflammation. Acute liver failure ensues when the extent of cellular death exceeds the hepatic regenerative ability. There are two typical patterns of cellular death occurring in acute liver injury/failure: necrosis and apoptosis. Necrosis involves the depletion of adenosine triphosphate (ATP), leading to cellular swelling and lysis, followed by the release of cellular contents which may induce secondary inflammation (Kaplowitz 2000). In contrast, the process of apoptosis is highly regulated and ATP-dependent to minimize the inflammatory response and the release of cellular contents to extracellular spaces (Spencer and Sorger 2011). The intrinsic and extrinsic pathways are the two main routes that have been described in acute liver injury/failure-related pathways (Rutherford and Chung 2008). The intrinsic apoptosis pathway involves the release of cytochrome c from the mitochondria, leading to the cascade responses of the caspase family. The extrinsic apoptosis pathway is characterized by signaling from a class of death receptors on the cell surface which induces programmed cell death through a ligand–receptor complex-mediated response (Spencer and Sorger 2011). Traditionally, necrosis and apoptosis are considered as separate events in acute liver injury. However, increasing reports point out that they are alternative outcomes of the same initiating factor and signaling pathway, a cellular event known as necro-apoptosis. Therefore, during acute liver injury/failure, the process of necrosis, apoptosis, and necroapoptosis are pathophysiological key events that lead to the damage of liver cells.

Unlike acute liver injury, chronic liver injury often originates from a repeated liver damage that happens within a relatively long period. Regardless of the cause, hepatic chronic injury usually results in a healing response that ultimately replaces the normal liver structure with regenerative process in the form of fibrosis or cirrhosis. Along with the impaired balance between endogenous vasoconstrictor (e.g., endothelin) and vasodilator (e.g., nitric oxide), portal hypertension, and increased intrahepatic resistance may occur as a consequence of this regenerative remodeling (Minano and Garcia-Tsao 2010). The prevalence of chronic liver injury continues to rise in recent years. According to a report from Centers for Disease Control and Prevention (CDC) of the USA, chronic liver injury and cirrhosis were the twelfth leading cause of mortality in the USA in 2007 (Xu et al. 2010). Compared to acute liver injury, chronic liver injury is induced by a broader spectrum of causes which can be roughly divided into five categories: (1) toxin- and drug-induced chronic liver injury (e.g., alcohol, amiodarone, and methotrexate; Aithal et al. 2011; Gao and Bataller 2011), (2) viral infection induced chronic liver injury (e.g., hepatitis B and hepatitis C; Thursz et al. 2011), (3) metabolic disorder induced chronic liver injury (e.g., nonalcoholic fatty liver disease and Wilson's disease; Nobili et al. 2011; Purchase 2013), (4) autoimmune-related chronic liver injury (e.g., autoimmune chronic hepatitis; Mayo 2011), and (5) other reasons (e.g., right heart failure induced chronic liver disease; Moller et al. 2009).

Several pathological events that occur during the development of chronic liver diseases which are similar to those of acute liver injury, including oxidative stress, inflammation, chemoattraction, necrosis, and programmed cell death (Quattroni 2011; Solis Herruzo et al. 2011). However, there are kinds of processes different from common acute injury of the liver, such as steatosis (hepatic accumulation of fatty droplets) and fibrosis/cirrhosis. Under normal conditions, the lipid metabolism in the liver must be kept in a balanced status. That is, a balance between lipid input (e.g., lipoprotein imported by low-density lipoprotein receptor and *de novo* lipogenesis) and lipid output (e.g., β-oxidation and production of very low-density lipoprotein). However, when the imbalance of such processes happens, elevated levels of free fatty acids (FFAs) and triglycerides can be observed in the liver. Three major mechanisms have been demonstrated in the production of excessive FFAs and triglycerides: (1) dietary FFAs imported in the liver through the uptake of intestinally derived chylomicron remnants, (2) FFAs from adipose tissue and diet that entered the liver through plasma nonesterified fatty acids (NEFAs) pool, and (3) locally synthesized fatty acids in the liver (hepatic *de novo* lipogenesis; Donnelly et al. 2005). Imbalance of hepatic lipid metabolism is one of the major causes for the pathogenesis of hepatic steatosis and insulin resistance (Tiniakos et al. 2010).

Liver fibrosis is a process of excessive accumulation of extracellular matrix (ECM) proteins (e.g., collagen) in the liver. It occurs in most types of chronic liver diseases, including viral hepatitis, alcohol abuse, nonalcoholic fatty liver disease (NAFLD), and even chronic CCl_4 treatment (Bataller and Brenner 2005; Schuppan and Kim 2013). The accumulation of ECM proteins damages the original hepatic architecture. A severe form of liver fibrosis may induce cirrhosis, an irreversible form of chronic liver disease, which eventually progresses to hepatocellular carcinoma (Schuppan and Afdhal 2008). Following hepatic inflammatory response, oxidative stress, and other forms of damage, hepatic stellate cells (HSCs) transform to myofibroblast-like cells to produce ECM proteins (Lotersztajn et al. 2005). Besides HSCs, portal myofibroblast cells and bone marrow-origin cells also participate in the fibrotic processes (Ramadori and Saile 2004). However, till now, activation of HSCs is considered as the central step towards fibrosis. In the normal liver, HSCs are in a quiescent form (as the storage place of vitamin A). The initiation process is probably caused by the paracrine actions from injured adjacent cell types, which include hepatocytes, Kupffer cells, sinusoidal endothelial cells, platelets, and infiltrated inflammatory cells (Maher 1999). Some cytokines and actins produced by neighboring injured cells can be used to activate HSCs as well. These include α-SMA, TGF-β_1, platelet-derived growth factor (PDGF), and endothelin (ET) (Gabele et al. 2005). Interestingly, activated HSCs can perpetuate their own activation by autocrine actions, such as the upregulation of TGFβ$_1$ receptor, as well as the increase of TGF-β_1 production (Friedman and Arthur 1989). Therefore, TGF-β_1 has been an important therapeutic option for the inhibition of HSCs activation (i.e., antagonism of TGF-β_1 activity).

Among various chronic liver disorders, fatty liver diseases, both alcoholic and nonalcoholic, received massive attention in recent years because they are affecting a great number of people in both affluent and developing countries. In this chapter, we will mainly focus on the etiology of fatty liver diseases and the protective effects of wolfberry on the pathogenesis of chronic liver diseases, with special emphasis at the molecular level.

3.2 Fatty Liver Diseases

3.2.1 *Definitions*

Fatty liver or fatty liver disease (FLD) refers to a clinical condition where vacuoles of triglyceride fat accumulate in liver cells (steatosis). Usually, FLD can be classified into two major categories: alcoholic fatty liver disease (AFLD) and nonalcoholic fatty liver disease (NAFLD), wherein the difference lies in the abuse consumption of alcohol for the former (over 30 g alcohol/day for men and 20 g alcohol/day for women; Alkhouri et al. 2009). They share similar histopathology but a different etiology and epidemiology (Scaglioni et al. 2011). It is not necessary to receive medical intervention during simple steatosis stage without hepatic inflammation and fibrosis, which can be reversed through the change of lifestyle (e.g., abstinence and weight reduction). However, when steatosis progresses to advanced alcoholic/nonalcoholic steatohepatitis (ASH and NASH, respectively), appropriate medical treatments are required to attenuate or retard the pathological progression, since a certain number of advanced fatty liver patients have the inherent propensity to progress toward cirrhosis and liver cancer. The latter conditions are life threatening (Reddy and Rao 2006). Besides, AFLD/NAFLD is now recognized to be closely associated with the metabolic syndrome (Marchesini et al. 2001), chronic hepatitis C (Moucari et al. 2008), hemochromatosis (Powell et al. 2005), and cardiovascular diseases (Ahmed et al. 2012). Therefore, it is vital to control the disease when it is still reversible.

3.2.2 *Epidemiology*

According to several recent surveys using ultrasonography, the prevalence of AFLD/NAFLD in the general population is 20–30 % in the Middle East (Zelber-Sagi et al. 2006) and Europe (Bedogni et al. 2005) and 15 % in China (Fan et al. 2005). For AFLD, it is probably the main cause of death among people with severe alcohol abuse and is responsible for about 3.8 % of global mortality (The World Health Report 2002). In general, up to 90 % of alcoholics have fatty liver and 5–15 % of them will develop ASH and cirrhosis over 20 years (Corrao et al. 1997). Simple steatosis is usually asymptomatic and may reverse after 4–6 weeks of abstinence from alcohol (Mendenhall 1968; Altamirano and Bataller 2011). Progression to fibrosis and cirrhosis may occur, however, in 5–15 % of patients despite abstinence from alcohol (Sorensen et al. 1984). It has been reported that females are at a greater risk than males for AFLD since women develop severe FLD and cirrhosis at a lesser intake of alcohol and fewer years of exposure (Nanji et al. 2002; Yokoyama et al. 2005). In a Japanese report, female patients showed a significant predominance of alcoholic liver disorders in histologically matched patient group with cumulative alcohol intake less than 500 kg for alcoholic hepatitis, less than 600 kg for hepatic fibrosis and less than 800 kg for liver cirrhosis (Hisatomi et al. 1997). Another 12-year prospective Danish study found that in over 13,000 participants, the risk of alcoholic liver diseases increases in women who consumed 84–156 g alcohol/week as compared with men who consumed 168–324 g alcohol/week (Becker et al. 1996). In addition, the development rate of alcohol-induced cirrhosis is higher in African–Americans and Hispanics as compared to Caucasians (Stinson et al. 2001).

For NAFLD, it is found in 70 % of the obese people and 35 % in lean individuals (Wanless and Lentz 1990). In the Western world, NAFLD influences 20–35 % of adults and 5–17 % of children (Browning et al. 2004; Schwimmer et al. 2006). In China, the prevalence of NAFLD is around 15 % (Fan and Farrell 2009), while Hong Kong shows a higher percentage about 27.5 % (Wong et al. 2012). The current gold standard for the diagnosis of NAFLD is liver biopsy. It is invasive, highly accurate but not realistic to perform in population-based studies. However, a Korean study using liver biopsy in donors for liver transplantation found that the prevalence of NAFLD is 51 % in 589 potential donors (Lee et al. 2007). A similar study in the USA has around 20 % of the liver donors with steatosis (Marcos et al. 2000). Another method to conduct histology-based judgment of NAFLD is autopsy, which showed similar NAFLD prevalence with ultrasonography and liver biopsy-based studies. In an Indian study of 1230 adult autopsies, there are 195 (16 %) cases with NAFLD. Only 5 % of them with high alcohol consumption history (Amarapurkar and Ghansar 2007). A recent report from Greece indicated a higher prevalence (31.3 %) of steatosis in 498 cases who died from ischemic heart disease or traffic accident after exclusion of hepatitis B or known liver diseases (Zois et al. 2010). It is well known that NAFLD shows gender and age differences in normal population. According to a recent survey in Japan, concerning cirrhotic NASH patients, the prevalence in women (~57 %) is higher than that in men (~43 %; Yatsuji et al. 2009). However, another survey which investigated the prevalence of HCC in already established NASH patients found that men exhibit higher prevalence (~62 %) when compared with women prevalence (~38 %; Hashimoto et al. 2009). The causes of gender differentiation of NAFLD may be attributed to estrogenic endocrinology, fat distribution, and lifestyle differences (e.g., smoking population ratio is higher in men than in women). However, detailed explanations and mechanisms of gender differentiation remain one of the most unresolved questions in the field of hepatology. Prevalence of NAFLD is also influenced by ethnicity. African–Americans show significantly less hepatic steatosis, although there is relatively high prevalence of obesity and diabetes (Caldwell et al. 2002). Hispanic Americans exhibit a high prevalence of steatosis, while Asian–Americans only show an intermediate level of prevalence of steatosis (Browning et al. 2004; Weston et al. 2005). Another study found that liver dysfunction in Japanese with severe obesity (BMI >35 kg/m^2) tends to be more severe than that in non-Japanese patients (Kakizaki et al. 2008). Difference of NAFLD prevalence is associated with different visceral adiposity and metabolic responses.

3.2.3   Etiology

It is indisputable that alcohol consumption is the leading cause of AFLD and subsequent ASH and cirrhosis. Persons who drink heavily (50–60 g of ethanol daily) represent a population at a higher risk for developing alcoholic liver disease (Bellentani et al. 1997; Kamper-Jorgensen et al. 2004). It is very interesting that despite their increased risk, the absolute risk of acquiring ASH or AFLD-induced cirrhosis is very low. For example, in three prospective studies, a total number of 1652 alcoholics who drank over 35 drinks in a week only had an absolute risk of alcoholic liver disorders of 5.9 % (Becker et al. 2002). In addition, drinking pattern also influences the chance to get AFLD. In a report from China, it was demonstrated that in a group of 1270 persons, those who drank wine outside mealtime showed 2.7 times more often than those drank only at mealtime (Lu et al. 2004). Besides the gender and ethnicity factors we discussed in the previous section, comorbid conditions significantly affect the likelihood of developing AFLD, including viral hepatitis (Bellentani et al. 1999), hepatic iron overload (Ioannou et al. 2004), obesity (Diehl 2004), gene polymorphisms (Tanaka et al. 1997), and environmental factors (Vidali et al. 2003).

To date, we still do not clearly know the exact etiology of NAFLD. It is generally agreed that insulin resistance and dysregulated lipid metabolism are responsible for the initiation of NAFLD (Utzschneider and Kahn 2006). Also, obesity is strongly associated with NAFLD, since most patients with NAFLD are obese and there is increasing evidence of an obesity epidemic in the USA and other countries (James et al. 2004; Flegal et al. 1998). Recent studies found that in the nonobese, the prevalence of NAFLD is increasing due to a spectrum of risk factors, including genetic background, certain lifestyle, environmental, and ethnic factors (Liu 2012).

3.2.4   Molecular Mechanisms

The pathogenesis of both AFLD and NAFLD shares several similarities at different metabolic levels. These mechanisms include increased lipid genesis and mobilization, as well as impaired lipid utilization and export from the liver (Zambo et al. 2013). Indeed, dysregulated hepatic immune function also greatly contributes to the progression of FLD (Feldstein et al. 2004). A "multi-hit" or "multiple parallel hits" theory has been proposed as a rough overview of the molecular mechanisms of FLD initiation and progression (Tilg and Moschen 2010). In this hypothesis, insulin resistance, together with other pathological factors (e.g., alcohol, genetic mutations, excess carbohydrates, and drug abuse), leads to the occurrence of fat accumulation (steatosis). After that, the liver is more vulnerable to following "multi-hit" such as oxidative stress, inflammation, chemoattraction, and necroapoptosis (steatohepatitis). Recent studies also demonstrated the important role of alcohol consumption causing an imbalance between lipogenic genes (e.g., sterol regulatory element-binding protein, SREBP) and lipolytic genes (e.g., peroxisome proliferator-activated receptor alpha, PPARα). Ethanol may also regulate adenosine 5'-monophosphate-activated protein kinase (AMPK) to influence the relative concentrations of intracellular malonyl coenzyme A (Malonyl-CoA) and long-chain acyl-coenzyme A, the key metabolites responsible for the balance between fat synthesis versus degradation pathways (Lakshman 2004). Clinically, patients with alcoholic liver disorders usually have circulating autoantibodies, hypergammaglobulinemia, antibodies to unique hepatic proteins, and cytotoxic lymphocytes reacting against autologous hepatocytes. These symptoms strongly reflect an altered immune regulation with the loss of immunotolerance (Thiele et al. 2004). The vital role of oxidative stress in alcohol hepatotoxicity has been established in recent years. Both alcohol exposure and cellular stresses induce the production of hepatic reactive oxygen species (ROS). ROS appears to activate AMPK signaling system, which has emerged in recent years as a kinase that controls the redox-state and mitochondrial function (Sid et al. 2013). Moreover, ROS can induce lipid peroxidation and inflammatory responses in the liver, resulting in steatohepatitis through cytochrome P450 family enzymes (De Minicis and Brenner 2008).

The molecular mechanisms of NAFLD progression have been well documented. First of all, the occurrence of insulin resistance in the liver, characterized by impaired sensitivity to insulin action, aggravates peripheral insulin receptor (IR) and contributes to hepatic lipogenesis (Bugianesi et al. 2005a). Consequently, hyperinsulinemia breaks the balance between lipogenesis and lipolysis, leading to an increased level of both circulating and hepatic levels of FFAs through decreased apolipoprotein synthesis (Bugianesi et al. 2005b). This process involves the signal transduction through the IR pathway. When insulin binds to IR, it will activate the phosphorylation of insulin receptor substrates (IRS-1 and IRS-2), which are the main mediators of insulin signaling in the liver (Kumashiro et al. 2011). Activated IRS recruits signaling molecules containing Src homology-2 domains, including phosphatidylinositol 3-kinase (PI3K), growth factor receptor-bound protein 2 (Grb2), and SH2 domain-containing tyrosine phosphatase (Shp2), which regulate the metabolic effects of insulin in a range of physiological processes (Sesti et al. 2001). The main effector of IRS signaling is the protein kinase Akt. When activated, Akt inactivates glycogen synthase kinase 3 (GSK3) to affect the glucose metabolism in the liver (Lee and Kim 2007). Moreover, alteration of insulin signaling modulates several kinases and transcription factors, which are crucial for the occurrence of hepatic fibrosis, proliferation, inflammation, apoptosis, and autophagy, including c-Jun N-terminal kinase (JNK), protein kinase C (PKC), mitogen-activated protein kinase (MAPK), and nuclear factor kappa-B (NF-κB) (Gao et al. 2002; Lee et al. 2003; Greene et al. 2006; Liu et al. 2007). Similar to AFLD, production of oxidative stress is vital for the progression of NAFLD because the ROS imbalance triggers steatohepatitis by lipid peroxidation. In a clinical study, it is found that at least 40% patients have an abnormal structure of the mitochondria (Pessayre et al. 2001). Mitochondrial respiratory complexes with impaired function are able to increase the expression of tumor necrosis factor-alpha (TNF-α), which causes additional lipid peroxidation of mitochondrial membranes, further aggravating mitochondrial function and hepatic necro-inflammation (Schwabe and Brenner 2006).

Moreover, lipid peroxidation will also induce the expression of other cytokines, such as transforming growth factor-beta (TGF-β), interleukin (IL)-6 and IL-8. Since the TGF/SMAD pathway contributes to hepatic fibrosis, activation of TGF-β will further exacerbate the progression of NAFLD (Xiao et al. 2013a). Besides, high level of hepatic proinflammatory cytokine expression and NF-κB activity is capable of causing hepatic apoptosis through both intrinsic and extrinsic apoptotic pathways (Xiao et al. 2013b).

3.2.5 Treatments

For AFLD, abstinence is the most important therapeutic intervention for patients. It has been demonstrated to significantly improve clinical outcomes and even to reverse fatty liver (Pessione et al. 2003). Pharmacological treatment is usually applied in acute alcoholic hepatitis, in association with alcohol abstinence (Luca et al. 1997). Indeed, it should be based on the disease stage and treatment aim (Sougioultzis et al. 2005). Naltrexone and acamprosate are two kinds of common pharmacological strategies which can assist in reducing or eliminating alcohol intake in chronic heavy drinkers (Bouza et al. 2004). Another drug, disulfiram, is believed to have favorable outcomes for patients under supervised administration (Garbutt et al. 1999). In alcohol-dependent patients with liver cirrhosis, baclofen showed effective in promoting the outcomes of abstinence (Addolorato et al. 2007). Besides these, in the clinics, it has long been established that patients with AFLD are nearly all malnourished, in which the severity of disease correlates with the degree of malnutrition (Halsted 2004). Therefore, nutritional supplementation therapy is provided either through enteral or parenteral support, and it may improve the survival of AFLD patients (Lochs and Plauth 1999).

For NAFLD, two major therapeutic strategies for NAFLD treatment are employed in the current stage: (1) lifestyle interventions including weight reduction, dietary modification, and physical exercise and (2) pharmaceutical therapy (Xiao et al. 2013c). We also examined several kinds of alternative therapies using plant derivatives, including garlic (Xiao et al. 2013d, e), green tea (Xiao et al. 2014a), and wolfberry (Xiao et al. 2013b). These derivatives significantly attenuate steatosis, fibrosis, oxidative stress, inflammation, and apoptosis in the injured liver.

3.3 Wolfberry on Fatty Liver Diseases

As early as 1997 and 1999, Kim et al. reported three cerebrosides isolated from *Lycium chinense* fruits with hepato-protective properties. They significantly reduced the levels of glutamic pyruvic transaminase (GPT) and sorbitol dehydrogenase (SDH) released from CCl_4-intoxicated rat primary hepatocytes (Kim et al. 1997, 1999). Later, in 2005, Ha et al. found that *L. chinense* Miller (Solanaceae) fruit

(LFE) has protective roles in CCl_4-induced hepatotoxicity in rats partly through cytochrome P450 2E1 (CYP2E1) (Ha et al. 2005). A following study confirmed the results in a H_2O_2-induced Chang liver cell damage model (Zhang et al. 2010). These effects were furthered investigated in a mice CCl_4 hepatotoxicity model in which we found that oral pretreatment with 1 or 10 mg/kg *Lycium barbarum* polysaccharides (LBP) prevents hepatic necrosis, oxidative stress, and inflammation, as well as promotes liver regeneration partly through the modulation of NF-κB activity (Xiao et al. 2012). In a streptozotocin-induced diabetic rat model, Li demonstrated that LBP restores abnormal oxidative indices near normal level in the liver and kidneys (Li 2007). In a similar model, Ye et al. found the beneficial effects of the extracts from Cortex Lycii Radicis, the dried root bark of *Lycium Chinese* Mill or *L. barbarum* L., on insulin resistance and dysregulates lipid metabolism in obese-diabetic rats (Ye et al. 2008).

The function of wolfberry in liver cancer has been characterized by several groups in recent years. In human hepatoma QGY7703 cell line, LBP treatment significantly inhibits QGY7703 cell growth with cycle arrest in S phase and induces cellular apoptosis. The concentration of intracellular Ca^{2+} is increased since the distribution of calcium in cells is changed (Zhang et al. 2005). Chao et al. also reported the proliferation-suppressive and apoptosis-inductive properties of hot water-extracted *L. barbarum* (LBE) in rat (H-4-II-E) and human HCC (HA22T/VGH) cell lines (Chao et al. 2006). However, the detailed anticancer mechanism of LBP in the liver still needs to be further elucidated.

3.4 Alcoholic Fatty Liver Disease (AFLD)

In an alcohol-induced rat liver injury model, it was found that a 30-day treatment with 7 g/kg ethanol causes typical AFLD features, including worsened hepatic histology, increased serum aminotransferases level, lipid level, and hepatic oxidative stress. Cotreatment with 300 mg/kg LBP significantly ameliorates liver injury, prevents the progression of alcohol-induced fatty liver, and improves the antioxidant functions when compared with the ethanol group (Cheng and Kong 2011). However, this study did not analyze the active component of LBP and elucidate the signaling pathways that contributed to the hepato-protective effects of LBP on AFLD. In our recent study, we found that zeaxanthin dipalmitate (ZD), a carotenoid from LBP, could effectively attenuate typical AFLD symptoms, including reduction in rat body weight, accumulation of fat droplets, occurrence of oxidative stress, inflammation, chemoattractive responses, and hepatic apoptosis in the liver. The attenuation of liver function abnormalities by ZD is partly through reduction in the level of CYP2E1, suppression of the elevated activity of NF-κB through the restoration of its inhibitor kappa B alpha (IκBα), and the modulation of MAPK pathways including p38 MAPK, JNK, and ERK. In addition, in the cellular AFLD model, we confirmed that the inhibition of p38 MAPK and ERK abolishes the beneficial effects of ZD (Xiao et al. 2014b). We also found that treatment with LBP reduces cellular oxidative stress and inflammation through inhibition of NLRP3 inflammasome in a thioredoxin-interacting protein (TXNIP)-dependent manner (Xiao et al. 2014c).

3.5   Non-alcoholic Fatty Liver Disease (NAFLD)

The first study regarding the ameliorative roles of wolfberry in NAFLD therapy was published in 2010. In this study, Kang et al. reported that after 13-week feeding with high-fat diet, another 8-week high-fat diet treatment with 5 or 10 % *L. chinense* leaf powder effectively reduces rat body weight, levels of serum triglyceride, and LDL-cholesterol with antioxidant effects (Kang et al. 2010). Another study found that both *L. barbarum* aqueous and ethanol extracts could significantly reduce liver damage and oxidative changes caused by an 8-week high-fat diet treatment, and brought back the antioxidants and lipids towards normal levels (Cui et al. 2011). These findings were further confirmed by Cho et al. since they indicated that tyramine derivatives from Lycii Cortex Radicis lowers the level of liver cholesterol and serum TBARS in high-fat diet-induced obese rats (Cho et al. 2011). We then demonstrated the antiobesity, antioxidant, and anti-inflammatory properties of LBP in a rat NAFLD model. Cotreatment with 1 mg/kg LBP and high-fat diet in an induced NAFLD rat model significantly alleviates hepatic injury, corrected lipid dysregulation, ameliorated fibrosis, oxidative stress, inflammation, and apoptosis (Xiao et al. 2013b). In our latest study, we induced NASH in a rat model by voluntary oral feeding of a high-fat diet ad libitum for 8 weeks. After that, 1 mg/kg LBP was administered for another 4 weeks with a high-fat diet. It is found that, when compared with NASH treatment for 12 weeks, therapeutic LBP treatment for 4 weeks after NASH induction exhibits ameliorative effects on: (1) increased body and wet liver weights, (2) hepatic insulin resistance and glucose metabolism dysfunction, (3) elevated level of serum aminotransferases, (4) fat accumulation in the liver and increased serum FFA level, (5) hepatic fibrosis, (6) hepatic oxidative stress, (7) hepatic inflammatory response, and (8) hepatic apoptosis. These improvements were partially through the modulation of transcription factor NF-κB, MAPK pathways, and the autophagic processes. In a palmitate acid-induced rat hepatocyte steatosis cell-based model, we also demonstrated that l-arabinose and β-carotene partially account for the beneficial effects of LBP on the hepatocyte (Xiao et al. 2014d).

3.6   Perspectives

In most published basic research and clinical studies using wolfberry, one of the major unsolved questions is the active component(s) from the extract. We have delineated that zeaxanthin part (Xiao et al. 2014b) and β-carotene (Xiao et al. 2014d) are partly responsible for the hepatic beneficial properties of wolfberry in animal models.

Admittedly, we cannot ignore the possible adverse effects brought by the application of wolfberry in the liver. Wang et al. reported that the extraction from *L. barbarum* could aggravate ischemia/reperfusion (I/R)-induced liver injury by increasing hydroxyl radical release with no effect on NO release (Wang et al. 2009). A recent case report from Spain indicated that a 60-year-old woman was admitted to the hospital because of asthenia, arthralgias, nonbloody diarrhea, and colic abdominal pain. It was found that she had a 10-day consumption of wolfberry tea three times a day but no other medication. In addition, most common causes of liver diseases (e.g., serum anti-HA–IgM, HbsAg, anti-HBc–IgM, anti-HCV, and nonorganspecific autoantibodies (ANA, ASMA, AMA, and anti-KLM1), CMV, EBV, HSV, VZV, IgG, IgA, IgM, alpha-1-fetoprotein, alpha-1-antitrypsin, serum iron and ferritin levels, and serum copper and ceruloplasmin levels were normal) were reasonably excluded. One month later, wolfberries were withdrawn and liver function tests returned to normal values (Arroyo-Martinez et al. 2011). Since the bioavailability and pharmacokinetic behavior of wolfberry, which contains more than 41 components, are still essentially unknown (Potterat 2010), we need to carefully evaluate its potential influence on human before launching a large-scale clinical trial.

References

Abdel-Misih SR, Bloomston M. Liver anatomy. Surg Clin North Am. 2010;90(4):643–53. doi:10.1016/j.suc.2010.04.017.

Addolorato G, Leggio L, Ferrulli A, Cardone S, Vonghia L, Mirijello A, Abenavoli L, D'Angelo C, Caputo F, Zambon A, Haber PS, Gasbarrini G. Effectiveness and safety of baclofen for maintenance of alcohol abstinence in alcohol-dependent patients with liver cirrhosis: randomised, double-blind controlled study. Lancet. 2007;370(9603):1915–22. doi:10.1016/s0140-6736(07)61814-5.

Ahmed MH, Barakat S, Almobarak AO. Nonalcoholic fatty liver disease and cardiovascular disease: has the time come for cardiologists to be hepatologists? J Obes. 2012;2012:483135. doi:10.1155/2012/483135.

Aithal GP, Watkins PB, Andrade RJ, Larrey D, Molokhia M, Takikawa H, Hunt CM, Wilke RA, Avigan M, Kaplowitz N, Bjornsson E, Daly AK. Case definition and phenotype standardization in drug-induced liver injury. Clin Pharmacol Ther. 2011;89(6):806–15. doi:10.1038/clpt.2011.58.

Alkhouri N, Lopez R, Berk M, Feldstein AE. Serum retinol-binding protein 4 levels in patients with nonalcoholic fatty liver disease. J Clin Gastroenterol. 2009;43(10):985–9. doi:10.1097/MCG.0b013e3181a0998d.

Altamirano J, Bataller R. Alcoholic liver disease: pathogenesis and new targets for therapy. Nat Rev Gastroenterol Hepatol. 2011;8:491–501. doi:10.1038/nrgastro.2011.134.

Amarapurkar A, Ghansar T. Fatty liver: experience from western India. Ann Hepatol. 2007;6(1):37–40.

Arroyo-Martinez Q, Saenz MJ, Arguelles Arias F, Acosta MS. Lycium barbarum: a new hepatotoxic "natural" agent? Dig Liver Dis. 2011;43(9):749. doi:10.1016/j.dld.2011.04.010.

Bataller R, Brenner DA. Liver fibrosis. J Clin Investig. 2005;115(2):209–18. doi:10.1172/jci24282.

Becker U, Deis A, Sorensen TI, Gronbaek M, Borch-Johnsen K, Muller CF, Schnohr P, Jensen G. Prediction of risk of liver disease by alcohol intake, sex, and age: a prospective population study. Hepatology (Baltimore, Md) 1996;23(5):1025–9. doi:10.1002/hep.510230513.

Becker U, Gronbaek M, Johansen D, Sorensen TI. Lower risk for alcohol-induced cirrhosis in wine drinkers. Hepatology (Baltimore, Md). 2002;35(4):868–75. doi:10.1053/jhep.2002.32101.

Bedogni G, Miglioli L, Masutti F, Tiribelli C, Marchesini G, Bellentani S. Prevalence of and risk factors for nonalcoholic fatty liver disease: the dionysos nutrition and liver study. Hepatology (Baltimore, Md). 2005;42(1):44–52. doi:10.1002/hep.20734.

Bellentani S, Saccoccio G, Costa G, Tiribelli C, Manenti F, Sodde M, Saveria Croce L, Sasso F, Pozzato G, Cristianini G, Brandi G. Drinking habits as cofactors of risk for alcohol induced liver damage. The dionysos study group. Gut. 1997;41(6):845–50.

Bellentani S, Pozzato G, Saccoccio G, Crovatto M, Croce LS, Mazzoran L, Masutti F, Cristianini G, Tiribelli C. Clinical course and risk factors of hepatitis C virus related liver disease in the general population: report from the Dionysos study. Gut. 1999;44(6):874–880.

Bouza C, Angeles M, Munoz A, Amate JM. Efficacy and safety of naltrexone and acamprosate in the treatment of alcohol dependence: a systematic review. Addiction (Abingdon, England). 2004;99(7):811–28. doi:10.1111/j.1360-0443.2004.00763.x.

Browning JD, Szczepaniak LS, Dobbins R, Nuremberg P, Horton JD, Cohen JC, Grundy SM, Hobbs HH. Prevalence of hepatic steatosis in an urban population in the United States: impact of ethnicity. Hepatology (Baltimore, Md). 2004;40(6):1387–95. doi:10.1002/hep.20466.

Bugianesi E, McCullough AJ, Marchesini G. Insulin resistance: a metabolic pathway to chronic liver disease. Hepatology (Baltimore, Md). 2005a;42 (5):987–1000. doi:10.1002/hep.20920.

Bugianesi E, Gastaldelli A, Vanni E, Gambino R, Cassader M, Baldi S, Ponti V, Pagano G, Ferrannini E, Rizzetto M. Insulin resistance in non-diabetic patients with non-alcoholic fatty liver disease: sites and mechanisms. Diabetologia. 2005b;48(4):634–42. doi:10.1007/s00125-005-1682-x.

Caldwell SH, Harris DM, Patrie JT, Hespenheide EE. Is NASH underdiagnosed among African Americans? Am J Gastroenterol. 2002;97(6):1496–500. doi:10.1111/j.1572-0241.2002.05795.x.

Chao JC, Chiang SW, Wang CC, Tsai YH, Wu MS. Hot water-extracted Lycium barbarum and Rehmannia glutinosa inhibit proliferation and induce apoptosis of hepatocellular carcinoma cells. World J Gastroenterol. 2006;12(28):4478–84.

Cheng D, Kong H. The effect of Lycium barbarum polysaccharide on alcohol-induced oxidative stress in rats. Molecules (Basel, Switzerland). 2011;16(3):2542–50. doi:10.3390/molecules16032542.

Cho SH, Park EJ, Kim EO, Choi SW. Study on the hypochlolesterolemic and antioxidative effects of tyramine derivatives from the root bark of Lycium chenese Miller. Nutr Res Pract. 2011;5(5):412–20. doi:10.4162/nrp.2011.5.5.412.

Corrao G, Ferrari P, Zambon A, Torchio P. Are the recent trends in liver cirrhosis mortality affected by the changes in alcohol consumption? Analysis of latency period in European countries. J Stud Alcohol. 1997;58(5):486–94.

Cui B, Liu S, Lin X, Wang J, Li S, Wang Q, Li S. Effects of Lycium barbarum aqueous and ethanol extracts on high-fat-diet induced oxidative stress in rat liver tissue. Molecules (Basel, Switzerland). 2011;16(11):9116–28. doi:10.3390/molecules16119116.

Czaja AJ, Freese DK. Diagnosis and treatment of autoimmune hepatitis. Hepatology (Baltimore, Md). 2002;36(2):479–97. doi:10.1053/jhep.2002.34944.

De Minicis S, Brenner DA. Oxidative stress in alcoholic liver disease: role of NADPH oxidase complex. J Gastroenterol Hepatol. 2008;23(Suppl 1):S98–103. doi:10.1111/j.1440-1746.2007.05277.x.

Diehl AM. Obesity and alcoholic liver disease. Alcohol. 2004;34(1):81–7. doi:http://dx.doi.org/10.1016/j.alcohol.2004.07.010.

Donnelly KL, Smith CI, Schwarzenberg SJ, Jessurun J, Boldt MD, Parks EJ. Sources of fatty acids stored in liver and secreted via lipoproteins in patients with nonalcoholic fatty liver disease. J Clin Investig. 2005;115(5):1343–51. doi:10.1172/jci23621.

Fan JG, Farrell GC. Epidemiology of non-alcoholic fatty liver disease in China. J Hepatol. 2009;50(1):204–10. doi:10.1016/j.jhep.2008.10.010.

Fan JG, Zhu J, Li XJ, Chen L, Li L, Dai F, Li F, Chen SY. Prevalence of and risk factors for fatty liver in a general population of Shanghai, China. J Hepatol. 2005;43 (3):508–14. doi:10.1016/j.jhep.2005.02.042.

Feldstein AE, Werneburg NW, Canbay A, Guicciardi ME, Bronk SF, Rydzewski R, Burgart LJ, Gores GJ. Free fatty acids promote hepatic lipotoxicity by stimulating TNF-alpha expression via a lysosomal pathway. Hepatology (Baltimore, Md). 2004;40(1):185–94. doi:10.1002/hep.20283.

Flegal KM, Carroll MD, Kuczmarski RJ, Johnson CL. Overweight and obesity in the United States: prevalence and trends, 1960–1994. Int J Obes Relat Metab Disord. 1998;22 (1):39–47.

Friedman SL, Arthur MJ. Activation of cultured rat hepatic lipocytes by Kupffer cell conditioned medium. Direct enhancement of matrix synthesis and stimulation of cell proliferation via induction of platelet-derived growth factor receptors. J Clin Investig. 1989;84(6):1780–5. doi:10.1172/jci114362.

Gabele E, Reif S, Tsukada S, Bataller R, Yata Y, Morris T, Schrum LW, Brenner DA, Rippe RA. The role of p70S6K in hepatic stellate cell collagen gene expression and cell proliferation. J Biol Chem. 2005;280(14):13374–82. doi:10.1074/jbc.M409444200.

Gao B, Bataller R. Alcoholic liver disease: pathogenesis and new therapeutic targets. Gastroenterology. 2011;141(5):1572–85. doi:10.1053/j.gastro.2011.09.002.

Gao Z, Hwang D, Bataille F, Lefevre M, York D, Quon MJ, Ye J. Serine phosphorylation of insulin receptor substrate 1 by inhibitor kappa B kinase complex. J Biological Chem. 2002;277 (50):48115–21. doi:10.1074/jbc.M209459200.

Garbutt JC, West SL, Carey TS, Lohr KN, Crews FT. Pharmacological treatment of alcohol dependence: a review of the evidence. JAMA. 1999;281(14):1318–25.

Gill RQ, Sterling RK. Acute liver failure. J Clin Gastroenterol. 2001;33(3):191–8.

Greene MW, Ruhoff MS, Roth RA, Kim JA, Quon MJ, Krause JA. PKCdelta-mediated IRS-1 Ser24 phosphorylation negatively regulates IRS-1 function. Biochem Biophys Res Commun. 2006;349(3):976–86. doi:10.1016/j.bbrc.2006.08.158.

Guilbert JJ.The world health report 2002—reducing risks, promoting healthy life. Educ Health (Abingdon). 2003;16(2):230.

Ha KT, Yoon SJ, Choi DY, Kim DW, Kim JK, Kim CH. Protective effect of Lycium chinense fruit on carbon tetrachloride-induced hepatotoxicity. J Ethnopharmacol. 2005;96(3):529–35. doi:10.1016/j.jep.2004.09.054.

Halsted CH. Nutrition and alcoholic liver disease. Semin Liver Dis. 2004;24(3):289–304. doi:10.1055/s-2004-832941.

Hashimoto E, Yatsuji S, Tobari M, Taniai M, Torii N, Tokushige K, Shiratori K. Hepatocellular carcinoma in patients with nonalcoholic steatohepatitis. J Gastroenterol. 2009;44(Suppl 19):89–95. doi:10.1007/s00535-008-2262-x.

Hisatomi S, Kumashiro R, Sata M, Ishii K, Tanikawa K. Gender difference in alcoholic liver disease in Japan—an analysis based on histological findings. Hepatol Res. 1997;8(2):113–20. doi:http://dx.doi.org/10.1016/S1386-6346(97)00062-4.

Horton JD, San Miguel FL, Membreno F, Wright F, Paima J, Foster P, Ortiz JA. Budd-Chiari syndrome: illustrated review of current management. Liver Int. 2008;28(4):455–66. doi:10.1111/j.1478-3231.2008.01684.x.

Ioannou GN, Dominitz JA, Weiss NS, Heagerty PJ, Kowdley KV. The effect of alcohol consumption on the prevalence of iron overload, iron deficiency, and iron deficiency anemia. Gastroenterology. 2004;126(5):1293–301.

Ishibashi H, Nakamura M, Komori A, Migita K, Shimoda S. Liver architecture, cell function, and disease. Semin Immunopathol. 2009;31(3):399–409. doi:10.1007/s00281-009-0155-6.

James PT, Rigby N, Leach R. The obesity epidemic, metabolic syndrome and future prevention strategies. Eur J Cardiovasc Prev Rehabil. 2004;11(1):3–8.

Kakizaki S, Takizawa D, Yamazaki Y, Nakajima Y, Ichikawa T, Sato K, Takagi H, Mori M, Kasama K. Nonalcoholic fatty liver disease in Japanese patients with severe obesity who received laparoscopic Roux-en-Y gastric bypass surgery (LRYGB) in comparison to non-Japanese patients. J Gastroenterol. 2008;43(1):86–92. doi:10.1007/s00535-007-2130-0.

Kamper-Jorgensen M, Gronbaek M, Tolstrup J, Becker U. Alcohol and cirrhosis: dose–response or threshold effect? J Hepatol 2004;41(1):25–30. doi:10.1016/j.jhep.2004.03.002.

Kang MH, Park WJ, Choi MK. Anti-obesity and hypolipidemic effects of Lycium chinense leaf powder in obese rats. J Med Food. 2010;13(4):801–7. doi:10.1089/jmf.2010.1032.

Kaplowitz N. Mechanisms of liver cell injury. J Hepatol. 2000;32(1 Suppl):39–47.

Kim SY, Choi YH, Huh H, Kim J, Kim YC, Lee HS. New antihepatotoxic cerebroside from Lycium chinense fruits. J Nat Prod. 1997;60(3):274–6. doi:10.1021/np960670b.

Kim SY, Lee EJ, Kim HP, Kim YC, Moon A, Kim YC. A novel cerebroside from lycii fructus preserves the hepatic glutathione redox system in primary cultures of rat hepatocytes. Biol Pharm Bull. 1999;22(8):873–5.

Korman JD, Volenberg I, Balko J, Webster J, Schiodt FV, Squires RH Jr, Fontana RJ, Lee WM, Schilsky ML. Screening for Wilson disease in acute liver failure: a comparison of currently available diagnostic tests. Hepatology (Baltimore, Md). 2008;48(4):1167–74. doi:10.1002/hep.22446.

Kumashiro N, Erion DM, Zhang D, Kahn M, Beddow SA, Chu X, Still CD, Gerhard GS, Han X, Dziura J, Petersen KF, Samuel VT, Shulman GI. Cellular mechanism of insulin resistance in nonalcoholic fatty liver disease. Proc Natl Acad Sci U S A. 2011;108(39):16381–5. doi:10.1073/pnas.1113359108.

Lakshman MR. Some novel insights into the pathogenesis of alcoholic steatosis. Alcohol. 2004;34(1):45–8. doi:10.1016/j.alcohol.2004.08.004.

Larson AM. Diagnosis and management of acute liver failure. Curr Opin Gastroenterol. 2010;26(3):214–21. doi:10.1097/MOG.0b013e32833847c5.

Larson AM, Polson J, Fontana RJ, Davern TJ, Lalani E, Hynan LS, Reisch JS, Schiodt FV, Ostapowicz G, Shakil AO, Lee WM. Acetaminophen-induced acute liver failure: results of a United States multicenter, prospective study. Hepatology (Baltimore, Md). 2005;42(6):1364–72. doi:10.1002/hep.20948.

Lee WM. Etiologies of acute liver failure. Semin Liver Dis. 2008;28(2):142–52. doi:10.1055/s-2008-1073114.

Lee J, Kim MS. The role of GSK3 in glucose homeostasis and the development of insulin resistance. Diabetes Res Clin Pract. 2007;77(Suppl 1):S49–57. doi:10.1016/j.diabres.2007.01.033.

Lee YH, Giraud J, Davis RJ, White MF. c-Jun N-terminal kinase (JNK) mediates feedback inhibition of the insulin signaling cascade. J Biol Chem. 2003;278(5):2896–902. doi:10.1074/jbc.M208359200.

Lee JY, Kim KM, Lee SG, Yu E, Lim YS, Lee HC, Chung YH, Lee YS, Suh DJ. Prevalence and risk factors of non-alcoholic fatty liver disease in potential living liver donors in Korea: a review of 589 consecutive liver biopsies in a single center. J Hepatol. 2007;47(2):239–44. doi:10.1016/j.jhep.2007.02.007.

Li XM. Protective effect of Lycium barbarum polysaccharides on streptozotocin-induced oxidative stress in rats. Int J Biol Macromol. 2007;40(5):461–5. doi:10.1016/j.ijbiomac.2006.11.002.

Liu CJ. Prevalence and risk factors for non-alcoholic fatty liver disease in Asian people who are not obese. J Gastroenterol Hepatol. 2012;27(10):1555–60. doi:10.1111/j.1440-1746.2012.07222.x.

Liu HY, Collins QF, Xiong Y, Moukdar F, Lupo EG Jr, Liu Z, Cao W. Prolonged treatment of primary hepatocytes with oleate induces insulin resistance through p38 mitogen-activated protein kinase. J Biol Chem. 2007;282(19):14205–12. doi:10.1074/jbc.M609701200.

Lochs H, Plauth M. Liver cirrhosis: rationale and modalities for nutritional support–the European Society of Parenteral and Enteral Nutrition consensus and beyond. Curr Opin Clin Nutr Metab Care. 1999;2(4):345–349.

Lotersztajn S, Julien B, Teixeira-Clerc F, Grenard P, Mallat A. Hepatic fibrosis: molecular mechanisms and drug targets. Ann Rev Pharmacol Toxicol. 2005;45:605–28. doi:10.1146/annurev.pharmtox.45.120403.095906.

Lu XL, Luo JY, Tao M, Gen Y, Zhao P, Zhao HL, Zhang XD, Dong N. Risk factors for alcoholic liver disease in China. World J Gastroenterol. 2004;10(16):2423–6.

Luca A, Garcia-Pagan JC, Bosch J, Feu F, Caballeria J, Groszmann RJ, Rodes J. Effects of ethanol consumption on hepatic hemodynamics in patients with alcoholic cirrhosis. Gastroenterology. 1997;112(4):1284–9.

Maher JJ. Leukocytes as modulators of stellate cell activation. Alcohol Clin Exp Res. 1999;23(5):917–21.

Marchesini G, Brizi M, Bianchi G, Tomassetti S, Bugianesi E, Lenzi M, McCullough AJ, Natale S, Forlani G, Melchionda N. Nonalcoholic fatty liver disease: a feature of the metabolic syndrome. Diabetes. 2001;50(8):1844–50.

Marcos A, Fisher RA, Ham JM, Olzinski AT, Shiffman ML, Sanyal AJ, Luketic VA, Sterling RK, Olbrisch ME, Posner MP. Selection and outcome of living donors for adult to adult right lobe transplantation. Transplantation. 2000;69(11):2410–5.

Mayo MJ. Management of autoimmune hepatitis. Curr Opin Gastroenterol. 2011;27(3):224–30. doi:10.1097/MOG.0b013e3283457ce0.

Mendenhall CL. Anabolic steroid therapy as an adjunct to diet in alcoholic hepatic steatosis. Am J Dig Dis. 1968;13(9):783–91.

Minano C, Garcia-Tsao G. Clinical pharmacology of portal hypertension. Gastroenterol Clin North Am. 2010;39(3):681–95. doi:10.1016/j.gtc.2010.08.015.

Miraglia R, Luca A, Gruttadauria S, Minervini MI, Vizzini G, Arcadipane A, Gridelli B. Contribution of transjugular liver biopsy in patients with the clinical presentation of acute liver failure. Cardiovasc Interv Radiol. 2006;29(6):1008–10. doi:10.1007/s00270-006-0052-5.

Moller S, Dumcke CW, Krag A. The heart and the liver. Expert Rev Gastroenterol Hepatol. 2009;3(1):51–64. doi:10.1586/17474124.3.1.51.

Moucari R, Asselah T, Cazals-Hatem D, Voitot H, Boyer N, Ripault MP, Sobesky R, Martinot-Peignoux M, Maylin S, Nicolas-Chanoine MH, Paradis V, Vidaud M, Valla D, Bedossa P, Marcellin P. Insulin resistance in chronic hepatitis C: association with genotypes 1 and 4, serum HCV RNA level, and liver fibrosis. Gastroenterology. 2008;134(2):416–23. doi:10.1053/j.gastro.2007.11.010.

Nanji AA, Su GL, Laposata M, French SW. Pathogenesis of alcoholic liver disease–recent advances. Alcohol Clin Exp Res. 2002;26(5):731–6.

Nobili V, Carter-Kent C, Feldstein AE. The role of lifestyle changes in the management of chronic liver disease. BMC Med. 2011;9:70. doi:10.1186/1741-7015-9-70.

O'Grady JG, Schalm SW, Williams R. Acute liver failure: redefining the syndromes. Lancet. 1993;342(8866):273–5.

Ostapowicz G, Fontana RJ, Schiodt FV, Larson A, Davern TJ, Han SH, McCashland TM, Shakil AO, Hay JE, Hynan L, Crippin JS, Blei AT, Samuel G, Reisch J, Lee WM. Results of a prospective study of acute liver failure at 17 tertiary care centers in the United States. Ann Intern Med. 2002;137(12):947–54.

Pessayre D, Berson A, Fromenty B, Mansouri A. Mitochondria in steatohepatitis. Semin Liver Dis. 2001;21(1):57–69.

Pessione F, Ramond MJ, Peters L, Pham BN, Batel P, Rueff B, Valla DC. Five-year survival predictive factors in patients with excessive alcohol intake and cirrhosis. Effect of alcoholic hepatitis, smoking and abstinence. Liver Int. 2003;23(1):45–53.

Polson J, Lee WM. AASLD position paper: the management of acute liver failure. Hepatology (Baltimore, Md). 2005;41(5):1179–97. doi:10.1002/hep.20703.

Potterat O. Goji (Lycium barbarum and L. chinense): phytochemistry, pharmacology and safety in the perspective of traditional uses and recent popularity. Planta Med. 2010;76(1):7–19. doi:10.1055/s-0029-1186218.

Powell EE, Ali A, Clouston AD, Dixon JL, Lincoln DJ, Purdie DM, Fletcher LM, Powell LW, Jonsson JR. Steatosis is a cofactor in liver injury in hemochromatosis. Gastroenterology. 2005;129 (6):1937–43. doi:10.1053/j.gastro.2005.09.015.

Purchase R. The treatment of Wilson's disease, a rare genetic disorder of copper metabolism. Sci Prog. 2013;96(Pt 1):19–32.

Quattroni P. Liver: usefulness of noninvasive biomarkers of fibrosis in chronic liver disease. Nat Rev Gastroenterol Hepatol. 2011;8(12):659. doi:10.1038/nrgastro.2011.200.

Ramadori G, Saile B. Portal tract fibrogenesis in the liver. Lab Invest. 2004;84(2):153–9. doi:10.1038/labinvest.3700030.

Reddy JK, Rao MS. Lipid metabolism and liver inflammation. II. Fatty liver disease and fatty acid oxidation. Am J Physiol Gastrointest Liver Physiol. 2006;290(5):G852–8. doi:10.1152/ajpgi.00521.2005.

Rutherford A, Chung RT. Acute liver failure: mechanisms of hepatocyte injury and regeneration. Semin Liver Dis. 2008;28(2):167–74. doi:10.1055/s-2008-1073116.

Scaglioni F, Ciccia S, Marino M, Bedogni G, Bellentani S. ASH and NASH. Dig Dis (Basel, Switzerland). 2011;29(2):202–10. doi:10.1159/000323886.

Schuppan D, Afdhal NH. Liver cirrhosis. Lancet. 2008;371(9615):838–51. doi:10.1016/s0140-6736(08)60383-9.

Schuppan D, Kim YO. Evolving therapies for liver fibrosis. J Clin Invest. 2013;123(5):1887–901. doi:10.1172/jci66028.

Schwabe RF, Brenner DA. Mechanisms of Liver injury. I. TNF-alpha-induced liver injury: role of IKK, JNK, and ROS pathways. Am J Physiol Gastrointest Liver Physiol. 2006;290(4):G583–9. doi:10.1152/ajpgi.00422.2005.

Schwimmer JB, Deutsch R, Kahen T, Lavine JE, Stanley C, Behling C. Prevalence of fatty liver in children and adolescents. Pediatrics. 2006;118(4):1388–93. doi:10.1542/peds.2006-1212.

Sesti G, Federici M, Hribal ML, Lauro D, Sbraccia P, Lauro R. Defects of the insulin receptor substrate (IRS) system in human metabolic disorders. FASEB J. 2001;15(12):2099–111. doi:10.1096/fj.01-0009rev.

Sid B, Verrax J, Calderon PB. Role of AMPK activation in oxidative cell damage: implications for alcohol-induced liver disease. Biochem Pharmacol. 2013;86(2):200–9. doi:10.1016/j.bcp.2013.05.007.

Solis Herruzo JA, Solis-Munoz P, Munoz Yague T, Garcia-Ruiz I. Molecular targets in the design of antifibrotic therapy in chronic liver disease. Rev Esp Enferm Dig. 2011;103(6):310–23.

Sorensen TI, Orholm M, Bentsen KD, Hoybye G, Eghoje K, Christoffersen P. Prospective evaluation of alcohol abuse and alcoholic liver injury in men as predictors of development of cirrhosis. Lancet. 1984;2(8397):241–4.

Sougioultzis S, Dalakas E, Hayes PC, Plevris JN. Alcoholic hepatitis: from pathogenesis to treatment. Curr Med Res Opin. 2005;21(9):1337–46. doi:10.1185/030079905X56493.

Spencer SL, Sorger PK. Measuring and modeling apoptosis in single cells. Cell. 2011;144(6):926–39. doi:10.1016/j.cell.2011.03.002.

Stinson FS, Grant BF, Dufour MC. The critical dimension of ethnicity in liver cirrhosis mortality statistics. Alcohol Clin Exp Res. 2001;25(8):1181–7.

Tanaka F, Shiratori Y, Yokosuka O, Imazeki F, Tsukada Y, Omata M. Polymorphism of alcohol-metabolizing genes affects drinking behavior and alcoholic liver disease in Japanese men. Alcohol Clin Exp Res. 1997;21(4):596–601.

The World Health Report—Reducing risks, promoting healthy life. 2002.

Thiele GM, Freeman TL, Klassen LW. Immunologic mechanisms of alcoholic liver injury. Semin Liver Dis. 2004;24(3):273–87. doi:10.1055/s-2004-832940.

Thursz M, Yee L, Khakoo S. Understanding the host genetics of chronic hepatitis B and C. Semin Liver Dis. 2011;31(2):115–27. doi:10.1055/s-0031-1276642.

Tilg H, Moschen AR. Evolution of inflammation in nonalcoholic fatty liver disease: the multiple parallel hits hypothesis. Hepatology (Baltimore, Md). 2010;52(5):1836–46. doi:10.1002/hep.24001.

Tiniakos DG, Vos MB, Brunt EM. Nonalcoholic fatty liver disease: pathology and pathogenesis. Ann Rev Pathol. 2010;5:145–71. doi:10.1146/annurev-pathol-121808-102132.

Utzschneider KM, Kahn SE. Review: the role of insulin resistance in nonalcoholic fatty liver disease. J Clin Endocrinol Metab. 2006;91(12):4753–61. doi:10.1210/jc.2006-0587.

Vidali M, Stewart SF, Rolla R, Daly AK, Chen Y, Mottaran E, Jones DE, Leathart JB, Day CP, Albano E. Genetic and epigenetic factors in autoimmune reactions toward cytochrome P4502E1 in alcoholic liver disease. Hepatology (Baltimore, Md). 2003;37(2):410–9. doi:10.1053/jhep.2003.50049.

Wang NT, Lin HI, Yeh DY, Chou TY, Chen CF, Leu FC, Wang D, Hu RT. Effects of the anti-oxidants lycium barbarum and ascorbic acid on reperfusion liver injury in rats. Transpl Proc. 2009;41(10):4110–3. doi:10.1016/j.transproceed.2009.08.051.

Wanless IR, Lentz JS. Fatty liver hepatitis (steatohepatitis) and obesity: an autopsy study with analysis of risk factors. Hepatology (Baltimore, Md). 1990;12(5):1106–10.

Weston SR, Leyden W, Murphy R, Bass NM, Bell BP, Manos MM, Terrault NA. Racial and ethnic distribution of nonalcoholic fatty liver in persons with newly diagnosed chronic liver disease. Hepatology (Baltimore, Md). 2005;41(2):372–9. doi:10.1002/hep.20554.

Wong VW, Chu WC, Wong GL, Chan RS, Chim AM, Ong A, Yeung DK, Yiu KK, Chu SH, Woo J, Chan FK, Chan HL. Prevalence of non-alcoholic fatty liver disease and advanced fibrosis in Hong Kong Chinese: a population study using proton-magnetic resonance spectroscopy and transient elastography. Gut. 2012;61(3):409–15. doi:10.1136/gutjnl-2011-300342.

Xiao J, Liong EC, Ching YP, Chang RC, So KF, Fung ML, Tipoe GL. Lycium barbarum polysaccharides protect mice liver from carbon tetrachloride-induced oxidative stress and necroinflammation. J Ethnopharmacol. 2012;139(2):462–70. doi:10.1016/j.jep.2011.11.033.

Xiao J, Ho CT, Liong EC, Nanji AA, Leung TM, Lau TY, Fung ML, Tipoe GL. Epigallocatechin gallate attenuates fibrosis, oxidative stress, and inflammation in non-alcoholic fatty liver disease rat model through TGF/SMAD, PI3K/Akt/FoxO1, and NF-kappa B pathways. Eur J Nutr. 2013a. doi:10.1007/s00394-013-0516-8.

Xiao J, Liong EC, Ching YP, Chang RC, Fung ML, Xu AM, So KF, Tipoe GL. Lycium barbarum polysaccharides protect rat liver from non-alcoholic steatohepatitis-induced injury. Nutr Diabetes. 2013b;3:e81. doi:10.1038/nutd.2013.22.

Xiao J, Guo R, Fung ML, Liong EC, Tipoe GL. Therapeutic approaches to non-alcoholic fatty liver disease: past achievements and future challenges. Hepatobiliary Pancreat Dis Int. 2013c;12(2):125–35.

Xiao J, Ching YP, Liong EC, Nanji AA, Fung ML, Tipoe GL. Garlic-derived S-allylmercaptocysteine is a hepato-protective agent in non-alcoholic fatty liver disease in vivo animal model. Eur J Nutr. 2013d;52(1):179–91. doi:10.1007/s00394-012-0301-0.

Xiao J, Guo R, Fung ML, Liong EC, Chang RC, Ching YP, Tipoe GL. Garlic-derived S-Allylmercaptocysteine ameliorates nonalcoholic fatty liver disease in a rat model through inhibition of apoptosis and enhancing autophagy. Evid Based Complement Alternat Med. 2013e;2013:642920. doi:10.1155/2013/642920.

Xiao J, Ho CT, Liong EC, Nanji AA, Leung TM, Lau TY, Fung ML, Tipoe GL. Epigallocatechin gallate attenuates fibrosis, oxidative stress, and inflammation in non-alcoholic fatty liver disease rat model through TGF/SMAD, PI3K/Akt/FoxO1, and NF-kappa B pathways. Eur J Nutr. 2014a;53(1):187–99. doi:10.1007/s00394-013-0516-8.

Xiao J, Wang J, Xing F, Han T, Jiao R, Liong EC, Fung ML, So KF, Tipoe GL. Zeaxanthin dipalmitate therapeutically improves hepatic functions in an alcoholic fatty liver disease model through modulating MAPK pathway. PloS ONE. 2014b;9(4):e95214. doi:10.1371/journal.pone.0095214.

Xiao J, Zhu Y, Liu Y, Tipoe GL, Xing F, So KF. Lycium barbarum polysaccharide attenuates alcoholic cellular injury through TXNIP-NLRP3 inflammasome pathway. Int J Biol Macromol. 2014c. 69:73–38. doi:10.1016/j.ijbiomac.2014.05.034.

Xiao J, Xing F, Huo J, Fung ML, Liong EC, Ching YP, Xu A, Chang RC, So KF, Tipoe GL. Lycium barbarum polysaccharides therapeutically improve hepatic functions in non-alcoholic steatohepatitis rats and cellular steatosis model. Sci Rep. 2014d(4): 5587 doi: 10.1038/srep05587.

Xu J, Kochanek K, Murphy S, Tejada-Vera B. Deaths: final data for 2007. National, vital statistics reports, Vol. 58. Hyattsville: National Center for Health Statistics; 2010.

Yatsuji S, Hashimoto E, Tobari M, Taniai M, Tokushige K, Shiratori K. Clinical features and outcomes of cirrhosis due to non-alcoholic steatohepatitis compared with cirrhosis caused by chronic hepatitis C. J Gastroenterol Hepatol. 2009;24(2):248–54. doi:10.1111/j.1440-1746.2008.05640.x.

Ye Z, Huang Q, Ni HX, Wang D. Cortex Lycii Radicis extracts improve insulin resistance and lipid metabolism in obese-diabetic rats. Phytother Res. 2008;22(12):1665–70. doi:10.1002/ptr.2552.

Yokoyama Y, Nimura Y, Nagino M, Bland KI, Chaudry IH. Current understanding of gender dimorphism in hepatic pathophysiology. J Surg Res. 2005;128(1):147–56. doi:10.1016/j.jss.2005.04.017.

Zambo V, Simon-Szabo L, Szelenyi P, Kereszturi E, Banhegyi G, Csala M. Lipotoxicity in the liver. World J Hepatol. 2013;5(10):550–7. doi:10.4254/wjh.v5.i10.550.

Zelber-Sagi S, Nitzan-Kaluski D, Halpern Z, Oren R. Prevalence of primary non-alcoholic fatty liver disease in a population-based study and its association with biochemical and anthropometric measures. Liver Int. 2006;26(7):856–63. doi:10.1111/j.1478-3231.2006.01311.x.

Zhang M, Chen H, Huang J, Li Z, Zhu C, Zhang S. Effect of lycium barbarum polysaccharide on human hepatoma QGY7703 cells: inhibition of proliferation and induction of apoptosis. Life Sci. 2005;76(18):2115–24. doi:10.1016/j.lfs.2004.11.009.

Zhang R, Kang KA, Piao MJ, Kim KC, Kim AD, Chae S, Park JS, Youn UJ, Hyun JW. Cytoprotective effect of the fruits of Lycium chinense Miller against oxidative stress-induced hepatotoxicity. J Ethnopharmacol. 2010;130(2):299–306. doi:10.1016/j.jep.2010.05.007.

Zois CD, Baltayiannis GH, Bekiari A, Goussia A, Karayiannis P, Doukas M, Demopoulos D, Mitsellou A, Vougiouklakis T, Mitsi V, Tsianos EV. Steatosis and steatohepatitis in postmortem material from Northwestern Greece. World J Gastroenterol. 2010;16(31):3944–9.

Chapter 4
Effects of *Lycium barbarum* on Modulation of Blood Vessel and Hemodynamics

Xue-Song Mi, Ruo-Jing Huang, Yong Ding, Raymond Chuen-Chung Chang and Kwok-Fai So

Abstract The blood-vessel-related mechanism plays an important role in the development of human diseases, e.g., brain degeneration, ischemic cardiomyopathy, diabetes, and tumors. In this chapter, the effects of *Lycium barbarum* (wolfberry) on blood vessel and hemodynamics will be discussed. Meanwhile, the related molecular and cellular mechanism will also be reviewed using current knowledge based on our studies and recent publications in other fields. The animal experiments showed that *L. barbarum* had protective effects on retinal blood vessels through the protection of endothelial cells and pericytes. The related mechanisms include decreasing the Immunoglobulin G (IgG) leakage and strengthening tight junctions (TJ) between endothelial cells. *L. barbarum* could modulate the response of blood-vessel-associated glial cells to impact on the integrity of blood-barriers. *L. barbarum* could inhibit the proliferation of smooth muscle cells under the high concentration of glucose. *L. barbarum* also showed effects on regulating the expression of blood vessel activating factors to modulate the vasoconstriction, dilatation, and neovascularization, such as endothelin-1 (ET-1), vascular endothelial growth factor (VEGF) and transforming growth factor-beta (TGF-β). To conclude, *L. barbarum* could remold the blood vessel and modulate hemodynamics by its effects on blood-vessel-related cells and activating factors.

K.-F. So (✉) · X.-S. Mi · R. C.-C. Chang
Department of Anatomy, The University of Hong Kong, Hong Kong, China
e-mail: hrmaskf@hku.hk

X.-S. Mi · R.-J. Huang · Y. Ding
Department of Ophthalmology, The First Affiliated Hospital of Jinan University, Guangzhou, Guangdong, China

K.-F. So · R. C.-C. Chang
GHM Institute of CNS Regeneration and Guangdong Key Laboratory of Brain Function and Diseases, Jinan University, Guangzhou, China

The State Key Laboratory of Brain and Cognitive Science and the Research Centre of Heart, Brain, Hormone and Healthy Aging, The University of Hong Kong, Hong Kong, China

K.-F. So
Department of Anatomy, The University of Hong Kong, Hong Kong, China

© Springer Science+Business Media Dordrecht 2015
R. C-C. Chang, K-F. So (eds.), *Lycium Barbarum and Human Health,*
DOI 10.1007/978-94-017-9658-3_4

Keywords *Lycium barbarum* · Neurovascular dysregulation · Endothelin-1 · Endothelial cell · Pericytes · Blood vessel

4.1 Introduction

The blood-vessel-related mechanism is referred to the change of blood vessel cells, blood-vessel-associated glial cells, and activating factors, resulting in the deregulation of microenvironment, which plays an important role in the development of human diseases.

The neurovascular unit consists of gather varieties of cells around the blood vessels, including vascular cells, glial cells, neurons, etc., which form a structural and functional integrity unit that maintains the balance of extravascular microenvironment (Carvey et al. 2009; Popescu et al. 2009; Iadecola 2010). Changes of microenvironment, such as blood-barrier permeability, or reduction of endothelial cells are detrimental to the survival of neurons in the retina and brain. The concept "vascular dysregulation" termed by Flammer in the mechanism of glaucoma, suggested that blood vessel damage could be a cause of neurodegeneration (Flammer 1994).

Lycium barbarum contains many bioactive components, lycium barbarum polysaccharides (LBP) is one of the major ingredients. Many studies analyzed the modulating effects of LBP on microvasculature. Since the stability of microenvironment is the guarantee for the survival of neurons. In studies of ocular hypertension insult, we investigated the vascular-related mechanisms of neurodegeneration in retinal ganglion cells (RGCs). Interestingly, we found that *L. barbarum* exhibits multiple vascular protective effects which enable it to be a potential traditional Chinese medicine targeting the complications or mechanisms of blood-vessel-related diseases.

The possible neuroprotective mechanism of LBP through the neurovascular unit, including proven effects of LBP on maintaining the integrality of blood vessel, down-regulating ET-1, amyloid beta (Aβ) and their related signal pathways, and rebuilding the structure and function of microenvironment (Mi et al. 2013; Yeh et al. 2008; He et al. 2012; Lau et al. 2012; Seeram 2011), as shown in a diagram (Fig. 4.1).

4.2 Maintaining the Integrality of Blood Vessels

The neurovascular unit consists of the endothelial cells, pericytes, astrocytes, microglia, and neurons, seen as Fig. 4.2. One function of neurovascular unit is to maintain the structural integrity of the blood–brain barriers (BBB) (Desai et al. 2007).

The BBB is critical in regulating the normal cerebral extracellular environment. Microvascular endothelium forms the first layer of BBB. The disruption of BBB resulted in the leakage of serum inflammatory protein from the circulation, which could destroy the endothelium and endothelial TJ (Farrall and Wardlaw 2009).

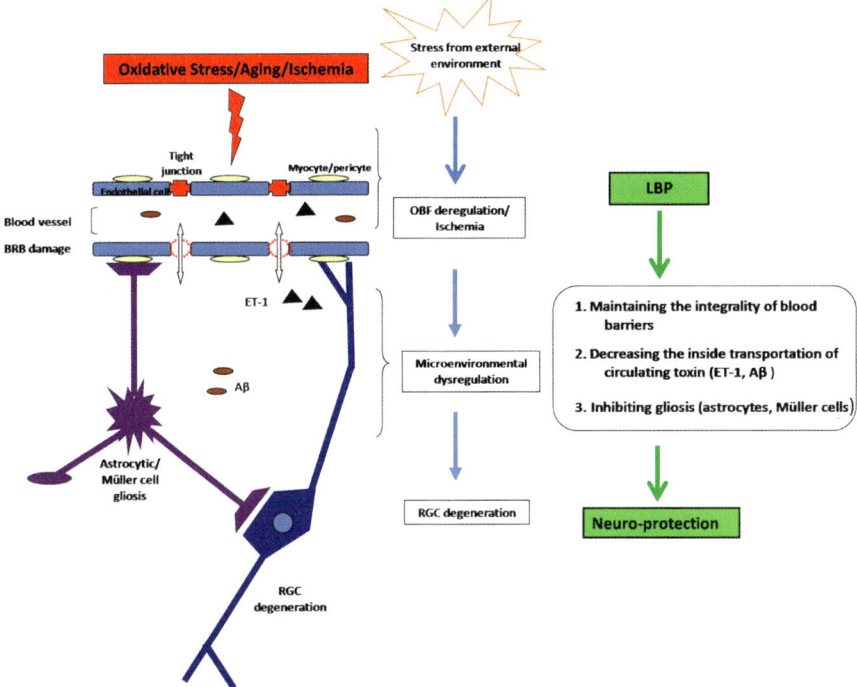

Fig. 4.1 Neurovascular dysregulation and neuroprotection mechanisms of LBP. Under the challenge of external insults, e.g., oxidative stress, aging, and ischemia, LBP inhibits the dysregulation of microenvironment and improves the survival of neurons. *LBP* polysaccharide of *Lycium barbarum*, *ET-1* endothelin-1, *Aβ* amyloid-β

Increased permeability of BBB can cause the intravascular fluid flowing into the brain parenchyma, resulting in perivascular edema. Edema is harmful to the supply of blood flow, contributing to the irreversible damage of brain (Kimelberg 1995). This forms a vicious cycle, which accelerates the degeneration of neurons.

The blood–retinal barrier (BRB) is an extension of the BBB. LBP could maintain the integrality of BBB and BRB, which showed in our previous studies (Mi et al. 2012b; Li et al. 2011). The effects of LBP on the protection of BBB include reducing Evans Blue (EB) extravasate, decreasing IgG leakage, upregulating the expression of occludin, and recover the integrity of BBB (Yang et al. 2012).

The BRB maintains physiological and structural balance within the retina, which contains inner and outer barriers. TJ between adjacent endothelial cells forms the inner BRB, while the outer barrier is formed by TJ between the retinal pigment epithelial (RPE) cells (Kim et al. 2013). As the extension of BBB, the permeability is a damage marker. To represent permeability, IgG leakage can be used to detect the damage of BRB (Cheung et al. 2005a). Our previous studies detected the protection

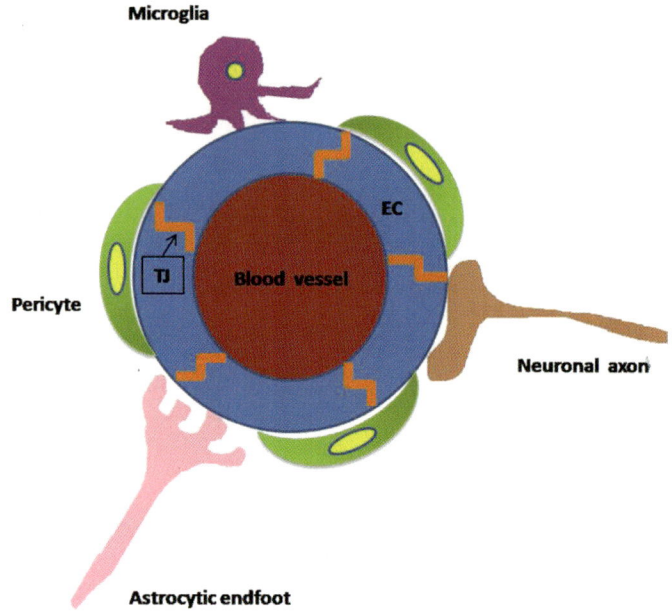

Fig. 4.2 The structure of neurovascular unit. In the central is the blood vessel. Around the blood vessel, there are neuronal axons, astrocytic endfeets and microglia, which cross-talk with blood vessel endothelial cells and pericytes. *EC* endothelial cell, *TJ* tight junction

of LBP on the BRB by decreasing IgG leakage (Fig. 4.3) (Mi et al. 2012b). A recent study indicated that LBP could increase the level of antiapoptotic protein Bcl-2 in epithelial cell (Song et al. 2012). RPE cells are particularly susceptible to oxidative stress (Beatty et al. 2000). Enhanced antiapoptotic protein Bcl-2 can act against oxidative stress and apoptosis (Ren et al. 2009). Thus, LBP could protect the BRB from disruption, not only for the inner but also for the outer barriers.

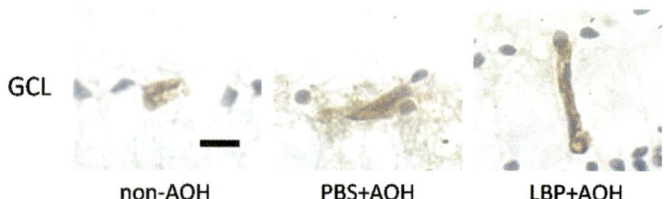

Fig. 4.3 Decreased leakage in LBP-fed-AOH retina at day 4 after AOH. This picture showed IgG leakages around blood vessels in AOH retina, while almost disappeared when pre-treated with LBP. Scale bar: 10 μm

4.3   Protection of Endothelial Cells and Pericytes

Endothelial cell and pericytes are main cell contents of blood vessel barriers. Pericytes are mural cells inside the blood vessel wall which are known to play key roles in regulating vascular formation, stability, function, and controlling endothelial proliferation (Hammes 2005). Under the challenge of stress from external environment, e.g., ischemia, the loss or proliferation of endothelial cells and pericytes will impact on the vascular remodeling.

Neovascularization is a kind of negative vascular remodeling, which will accrue under the condition of the loss or proliferation of endothelial cells and pericytes. Neovascularization is also a kind of blood vessel response for compensation of ischemia in tissue. However, it will become a vicious circle so that ischemia could induce the loss of endothelial cells and pericytes. In response, endothelial cells and pericytes will proliferate and aggravate the ischemia in tissue.

Thus, one finding of our previous study showed that LBP could prevent the loss of endothelial cells and pericytes at the early stage of ischemia (since in the experimental design, we fed the animal the LBP solution in advance of ischemia challenge), which played a role to prevent the disadvantaged change of vascular remodeling (Mi et al. 2012b).

4.4   Regulating the Response of Blood Vessel-Associated Astrocytes

4.4.1   Astrocytes

Astrocytes are divided into protoplasmic and fibrous types based on their morphologic characters and the distribution (Blackwood 1976; Hirano 1985; Privat and Rataboul 1987). The morphology of astrocytes are complex, different morphology of astrocytes can coexist in a given brain region (Matyash and Kettenmann 2010). Astrocytes can also express different functions in gap junction coupling, expression of transmitter receptors, membrane currents, and glutamate transporters (Matyash and Kettenmann 2010). The distribution of astrocytes is strongly determined by vasculature (Stone and Dreher 1987). Astrocytes which are closed to retinal vessels (Schnitzer 1987) will help maintain vascular integrity (Zhang and Stone 1997). Astrocytes enlarge their end-feet around blood vessels, participating in the formation of TJ of endothelial cells which could prevent BBB from damage, and also modulate the transport function of the cerebral endothelium (Montgomery 1994). Astrocytes are known to be the main target of ET-1 (Sasaki et al. 1997; Prasanna et al. 2003). Previous studies found that astrocytic ET-1 has deleterious effects on water homeostasis, cerebral edema, and BBB integrity, which contribute to severe ischemic brain injury (Lo et al. 2005). Disruption of BRB leads to the swelling of astrocytes associated with the overexpression of glial fibrillary acidic protein

(GFAP) and aquaporin-4 (AQP4) under the ischemic insult (Kaur et al. 2007). Each astrocyte could form end-feet around at least one blood vessel, while blood vessels can affect the morphology of astrocytes (Bushong et al. 2002). Our previous study showed that LBP can decrease the number of proliferated astrocytes and reduce their end-foot edema around the blood vessel, which could protect the survival of neurons. Thus, LBP can induce the blood-vessel-associated astrocytes to show entirely different morphologic change. These changes included apparently more processes touching blood vessels without enlargement of their end-feet (Leung et al. 2009b; Lo et al. 2005; Yeung et al. 2009).

4.4.2 Reducing the Expression of GFAP

In normal retina, GFAP is expressed on astrocytes and the end-feet of Müller cells. When under the stress, the expression of GFAP will increase not only on astrocytes but also on the process of the Müller cells (Cheung et al. 2005b; Mori et al. 2009; Taylor et al. 2010; Hirrlinger et al. 2010). It was reported that the overexpression of GFAP could be as the indicator for the reduction of blood flow (Osborne et al. 1991). From our previous studies, LBP could reduce the expression of GFAP, which may be a possible mechanism for recovery of blood supply to the tissue (Mi et al. 2012b).

4.4.3 Decreasing the Expression of AQP4

AQP4 is a member of water channel protein family which mainly expressed in astrocytes (Asai et al. 2013) and also a marker to reflect the capability to maintain the balance of water across the wall of blood vessels and to detect the functional change of astrocytes (Nagelhus et al. 1998). AQP4 is linked to the damage of BBB, suggesting the increase of permeability (Guo et al. 2006). It was reported that overexpression of AQP4 was related to the retinal swelling and involved in the process of neuronal death (Verkman et al. 2008). Previous studies demonstrated that pretreatment with LBP could reduce the expression of AQP4 in the animal model of ischemia in retina and brain (Li et al. 2011; Mi et al. 2012b; Yang et al. 2012).

4.5 Regulating Blood-Vessel-Related Factors and the Blood Flow

4.5.1 ET-1

ET-1, synthesized in vascular endothelial cells, is a potent vasoconstrictor. ET-1 regulates blood flow mainly via its two receptors, ET_A and ET_B (Narayan et al. 2003). ET_A is expressed on vascular smooth muscle cells and contracts blood vessels, while ET_B is mainly expressed in the endothelium and astrocytes and dilates blood vessels (Narayan et al. 2003; Masaki et al. 1994). Overexpression of ET-1 results in downregulating the level of occludin, the key protein located at TJ between blood vessel endothelial cells to maintain the integrity of blood-barrier (Leung et al. 2009a). Decreased expression of occludin has been reported in middle cerebral artery occlusion (MCAO)-induced ischemic brain and retina. However, LBP-pretreatment could maintain the level of occludin among blood vessels (Yang et al. 2012; Mi et al. 2012b). Further studies showed that LBP could reduce the ET-1-ET_A signal pathway while activate ET-1-ET_B pathway in the retinal blood vessels through downregulating the expression of ET_A and upregulating ET_B (Yorio et al. 2002). Not only endothelial cells, ET-1 is considered to stimulate contraction of pericytes (Ramachandran et al. 1993). Pericytes are considered to be other contractile cells to help regulating the blood flow (Wallow and Burnside 1980). Thus, through regulating the expression of ET_A and ET_B receptors, LBP could modulate the blood flow of retinal blood vessels (Mi et al. 2012a, b). Thus, LBP could protect endothelial cells and pericytes via downregulating ET-1.

4.5.2   Vascular Endothelial Growth Factor (VEGF) and Transforming Growth Factor-Beta (TGF-β)

Angiogenesis is the generation of new blood vessels from the original ones, playing an important role in the pathogenesis of diabetes. VEGF and TGF-β are angiogenin mediators. The high level of glucose in blood could result in the overexpression of VEGF (Martin et al. 2003) and TGF-β (Singh et al. 2013), contributing to vascular abnormalities in diabetes. LBP was shown to have a significant effect on reducing VEGF (Tian et al. 2013) and TGF-β (Singh et al. 2013). One recent study reported that a wolfberry contented formula diet (including *Rhizoma dioscorea*, *L. barbarum*, *Prunella vulgaris* and hawthorn) had effects on antidiabetes (Cheng et al. 2014).

4.5.3   The Receptor for Advanced Glycation Endproducts (RAGE)

RAGE is expressed on both smooth muscle cells and endothelial cells of blood vessel (Farmer and Kennedy 2009). It was reported that one role of RAGE was to transport Aβ across BBB to the brain tissue. Thereafter, Aβ could interact with RAGE resulting in the release of ET-1 from endothelial cells (Deane et al. 2003a), and contributing to vascular inflammation and endothelial dysfunction (Harja et al. 2008). Our previous study has proved that LBP can reduce the expression of RAGE (Mi et al. 2012b), therefore to inhibit the destructive effect of RAGE to blood vessels under ischemic insults.

4.5.4 Aβ

Aβ is a ligand of RAGE, which can induce vasoconstriction, decrease cerebral flow, and enhance the effect of ET-1-induced vasoconstriction (Paris et al. 2000). The accumulation of Aβ transported from BBB is also toxic to vasculature (Zlokovic 2005). ET-1 could inhibit the cerebral blood flow (CBF) through activating Aβ-RAGE transportation across the BBB (Deane et al. 2003b). LBP could downregulate the expression of Aβ (Mi et al. 2012b) thus to inhibit the Aβ-induced vascular toxic effect (Ho et al. 2010).

4.5.5 Advanced Glycation Endproducts (AGEs)

AGE is another key ligand of RAGE. The accumulation of AGE will activate RAGE, leading to endothelial dysfunction (Xu et al. 2003). Besides, AGE-RAGE pathway could enhance the synthesis of ET-1 in the endothelial cells (Quehenberger et al. 2000). Amount of evidences showed that AGE could induce the diabetic vascular complication (Takeuchi et al. 2000; Yamagishi et al. 2002b). AGE could induce pericyte apoptosis and VEGF expression contributing to the destruction of retinal vasculature (Yamagishi et al. 2002a). Thus, accumulation of AGE and over-expression of VEGF will aggravate the disorder of the vascular system (Yokoi et al. 2005). In our previous study, it was demonstrated that LBP could downregulate the expression of AGE, which could inhibit the destructive effect of AGE to blood vessels (Mi et al. 2012b).

4.6 Blood Flow and Hemodynamics

In normal microenvironment, the neurovascular unit regulates the blood flow by the end-feet activity of astrocytes (Iadecola and Nedergaard 2007), the autoregulation of blood vessels (Flammer and Mozaffarieh 2007) and vascular activating factors, such as ET-1 (Geijssen and Greve 1995). LBP was reported to dilate blood vessels and enhance the blood flow to the kidney in hypertensive rats (Jia et al. 1998). The mechanism is related to regulating the endothelin system.

The blood supply to the eye is referred to the difference between blood pressure and ocular pressure. Nocturnal systemic hypotension increases the risk of glaucoma progression (Charlson et al. 2014). In ocular hypertension, the experimental evidence showed that the capacity of blood flow autoregulation is vulnerable to the low blood pressure. This hemodynamic mechanism plays an important role in the development of glaucoma (Wang et al. 2014). Since the blood flow to optic nerve head is more easily observed to the decrease of the blood pressure (Wang et al. 2014). Although there was no direct evidence to show *L. barbarum* could increase ocular blood flow (OBF) to optic nerve head, through regulating endothelin and its receptors ETA and ETB, *L. barbarum* may be beneficial to the blood supply of retinal neurons.

In summary, from the above evidence, it suggests that *L. barbarum* is beneficial to maintain the microcirculation and sustain the blood flow through protecting the survival of blood vessel cells and regulating the response of blood-vessel-associated glial cells, together with mediating the vascular-related factors, e.g., ET-1, Aβ, RAGE and AGE which are all detrimental to the vasculature when the tissue was challenged by the external stress.

4.7 Perspectives

Vascular complication is an important pathogenesis and disability factor for many human diseases. It has also become a huge burden to health professions and finance of the whole society. As a vascular protective agent, *L. barbarum* may be promising to be used in prevention and treatment of vascular-related diseases.

Whereas more specific mechanisms and applications of *L. barbarum* in other areas of vascular complications are needed. Further investigation will be useful to translate experimental findings to clinical applications.

Conflicts of Interest None declared

References

Asai H, Kakita H, Aoyama M, Nagaya Y, Saitoh S, Asai K. Diclofenac enhances proinflammatory cytokine-induced aquaporin-4 expression in cultured astrocyte. Cell Mol Neurobiol. 2013;33(3):393–400.

Beatty S, Koh H-H, Phil M, Henson D, Boulton M. The role of oxidative stress in the pathogenesis of age-related macular degeneration. Surv Ophthalmol. 2000;45(2):115–34.

Blackwood W. Normal structure and general pathology of the nerve cell and neuroglia. In: Blackwood W, Corsellis JAN, editors. Greenfield's neuropathology. London: Edward Arnold; 1976.

Bushong EA, Martone ME, Jones YZ, Ellisman MH. Protoplasmic astrocytes in CA1 stratum radiatum occupy separate anatomical domains. J Neurosci. 2002;22(1):183–92.

Carvey PM, Hendey B, Monahan AJ. The blood-brain barrier in neurodegenerative disease: a rhetorical perspective. J Neurochem. 2009;111(2):291–314.

Charlson ME, de Moraes CG, Link A, Wells MT, Harmon G, Peterson JC, Ritch R, Liebmann JM. Nocturnal systemic hypotension increases the risk of glaucoma progression. Ophthalmology. 2014. doi:10.1016/j.ophtha.2014.04.016.

Cheng Q, Zhang X, Wang O, Liu J, Cai S, Wang R, Zhou F, Ji B. Anti-diabetic effects of the ethanol extract of a functional formula diet in mice fed with a fructose/fat-rich combination diet. J Sci Food Agric. 2014. doi:10.1002/jsfa.6737.

Cheung AK, Fung MK, Lo AC, Lam TT, So KF, Chung SS, Chung SK. Aldose reductase deficiency prevents diabetes-induced blood-retinal barrier breakdown, apoptosis, and glial reactivation in the retina of db/db mice. Diabetes. 2005a;54(11):3119–25.

Cheung AK, Fung MK, Lo AC, Lam TT, So KF, Chung SS, Chung SK. Aldose reductase deficiency prevents diabetes-induced blood-retinal barrier breakdown, apoptosis, and glial reactivation in the retina of db/db mice. Diabetes. 2005b;54(11):3119–25.

Deane R, Du Yan S, Submamaryan RK, LaRue B, Jovanovic S, Hogg E, Welch D, Manness L, Lin C, Yu J. RAGE mediates amyloid-β peptide transport across the blood-brain barrier and accumulation in brain. Nat Med. 2003a;9(7):907–13.

Deane R, Du Yan S, Submamaryan RK, LaRue B, Jovanovic S, Hogg E, Welch D, Manness L, Lin C, Yu J, Zhu H, Ghiso J, Frangione B, Stern A, Schmidt AM, Armstrong DL, Arnold B, Liliensiek B, Nawroth P, Hofman F, Kindy M, Stern D, Zlokovic B. RAGE mediates amyloid-beta peptide transport across the blood-brain barrier and accumulation in brain. Nat Med. 2003b;9(7):907–13. doi:10.1038/nm890.

Desai BS, Monahan AJ, Carvey PM, Hendey B. Bloodbrain barrier pathology in Alzheimer's and Parkinson's disease: implications for drug therapy. Cell Transplant. 2007;16(3):285–99.

Farmer DG, Kennedy S. RAGE, vascular tone and vascular disease. Pharmacol Ther. 2009;124(2):185–94.

Farrall AJ, Wardlaw JM. Blood-brain barrier: ageing and microvascular disease—systematic review and meta-analysis. Neurobiol Aging. 2009;30(3):337–52. doi:10.1016/j.neurobiolaging.2007.07.015.

Flammer J. The vascular concept of glaucoma. Surv Ophthalmol. 1994;38:38S3–6.

Flammer J, Mozaffarieh M. What is the present pathogenetic concept of glaucomatous optic neuropathy? Surv Ophthalmol. 2007;52(6):S162-73.

Geijssen H, Greve E. Vascular concepts in glaucoma. Curr Opin Ophthalmol. 1995;6(2):71–7.

Guo Q, Sayeed I, Baronne LM, Hoffman SW, Guennoun R, Stein DG. Progesterone administration modulates AQP4 expression and edema after traumatic brain injury in male rats. Exp Neurol. 2006;198(2):469–78.

Hammes H-P. Pericytes and the pathogenesis of diabetic retinopathy. Horm Metab Res. 2005;37(S1):39–43.

Harja E, Bu D-x, Hudson BI, Chang JS, Shen X, Hallam K, Kalea AZ, Lu Y, Rosario RH, Oruganti S. Vascular and inflammatory stresses mediate atherosclerosis via RAGE and its ligands in apoE-/-mice. J Clin Invest. 2008;118(1):183–94.

He N, Yang X, Jiao Y, Tian L, Zhao Y. Characterisation of antioxidant and antiproliferative acidic polysaccharides from Chinese wolfberry fruits. Food Chem. 2012;133(3):978–89.

Hirano A. Neurons, astrocytes, and ependyma. Textbook of neuropathology. Baltimore: Williams & Wilkins; 1985. pp. 1–91.

Hirrlinger PG, Ulbricht E, Iandiev I, Reichenbach A, Pannicke T. Alterations in protein expression and membrane properties during Müller cell gliosis in a murine model of transient retinal ischemia. Neurosci Lett. 2010;472(1):73–8.

Ho Y-S, Yu M-S, Yang X-F, So K-F, Yuen W-H, Chang RC-C. Neuroprotective effects of polysaccharides from wolfberry, the fruits of Lycium barbarum, against homocysteine-induced toxicity in rat cortical neurons. J Alzheimer's Dis. 2010;19(3):813–27.

Iadecola C. The overlap between neurodegenerative and vascular factors in the pathogenesis of dementia. Acta Neuropathol. 2010;120(3):287–96.

Iadecola C, Nedergaard M. Glial regulation of the cerebral microvasculature. Nat Neurosci. 2007;10(11):1369–76.

Jia YX, Dong JW, Wu XX, Ma TM, Shi AY. The effect of lycium barbarum polysaccharide on vascular tension in two-kidney, one clip model of hypertension. Sheng Li Xue Bao. 1998;50(3):309–14.

Kaur C, Sivakumar V, Yong Z, Lu J, Foulds W, Ling E. Blood-retinal barrier disruption and ultrastructural changes in the hypoxic retina in adult rats: the beneficial effect of melatonin administration. J Pathol. 2007;212(4):429–39.

Kim J, Kim CS, Lee IS, Lee YM, Sohn E, Jo K, Kim JH, Kim JS. Extract of Litsea japonica ameliorates blood-retinal barrier breakdown in db/db mice. Endocrine. 2013. doi:10.1007/s12020-013-0085-x.

Kimelberg HK. Current concepts of brain edema. Review of laboratory investigations. J Neurosurg. 1995;83(6):1051–9. doi:10.3171/jns.1995.83.6.1051.

Lau BW-M, Chia-Di Lee J, Li Y, Fung SM-Y, Sang Y-H, Shen J, Chang RC-C, So K-F. Polysaccharides from wolfberry prevents corticosterone-induced inhibition of sexual behavior and increases neurogenesis. PLoS ONE. 2012;7(4):e33374.

Leung JW, Chung SS, Chung SK. Endothelial endothelin-1 over-expression using receptor tyrosine kinase tie-1 promoter leads to more severe vascular permeability and blood brain barrier breakdown after transient middle cerebral artery occlusion. Brain Res. 2009a;1266:121–9. doi:10.1016/j.brainres.2009.01.070.

Leung JW, Chung SS, Chung SK. Endothelial endothelin-1 over-expression using receptor tyrosine kinase *tie-1* promoter leads to more severe vascular permeability and blood brain barrier breakdown after transient middle cerebral artery occlusion. Brain Res. 2009b;1266:121–9.

Li SY, Yang D, Yeung CM, Yu WY, Chang RC, So KF, Wong D, Lo AC. Lycium barbarum polysaccharides reduce neuronal damage, blood-retinal barrier disruption and oxidative stress in retinal ischemia/reperfusion injury. PLoS ONE. 2011;6(1):e16380. doi:10.1371/journal.pone.0016380.

Lo AC, Chen AY, Hung VK, Yaw LP, Fung MK, Ho MC, Tsang MC, Chung SS, Chung SK. Endothelin-1 overexpression leads to further water accumulation and brain edema after middle cerebral artery occlusion via aquaporin 4 expression in astrocytic end-feet. J Cereb Blood Flow Metab. 2005;25(8):998–1011.

Martin A, Komada MR, Sane DC. Abnormal angiogenesis in diabetes mellitus. Med Res Rev. 2003;23(2):117–45. doi:10.1002/med.10024.

Masaki T, Vane JR, Vanhoutte PM. International union of pharmacology nomenclature of endothelin receptors. Pharmacol Rev. 1994;46(2):137–42.

Matyash V, Kettenmann H. Heterogeneity in astrocyte morphology and physiology. Brain Res Rev. 2010;63(1):2–10.

Mi X-S, Chiu K, Van G, Leung J, Lo A, Chung SK, Chang R, So K-F. Effect of Lycium barbarum polysaccharides on the expression of endothelin-1 and its receptors in an ocular hypertension model of rat glaucoma. Neural Regen Res. 2012a;7(9):645.

Mi XS, Feng Q, Lo AC, Chang RC, Lin B, Chung SK, So KF. Protection of retinal ganglion cells and retinal vasculature by *Lycium barbarum* polysaccharides in a mouse model of acute ocular hypertension. PLoS ONE. 2012b;7(10):e45469. doi:10.1371/journal.pone.0045469.

Mi XS, Zhong JX, Chang RC, So KF. Research advances on the usage of traditional Chinese medicine for neuroprotection in glaucoma. J Integr Med. 2013;11(4):233–40. doi:10.3736/jintegrmed2013037.

Montgomery DL. Astrocytes: form, functions, and roles in disease. Vet Pathol. 1994;31(2):145–67.

Mori A, Saigo O, Hanada M, Nakahara T, Ishii K. Hyperglycemia accelerates impairment of vasodilator responses to acetylcholine of retinal blood vessels in rats. J Pharmacol Sci. 2009;110(2):160–8.

Nagelhus EA, Veruki ML, Torp R, Haug F-M, Laake JH, Nielsen S, Agre P, Ottersen OP. Aquaporin-4 water channel protein in the rat retina and optic nerve: polarized expression in Müller cells and fibrous astrocytes. J Neurosci. 1998;18(7):2506–19.

Narayan S, Prasanna G, Krishnamoorthy RR, Zhang X, Yorio T. Endothelin-1 synthesis and secretion in human retinal pigment epithelial cells (ARPE-19): differential regulation by cholinergics and TNF-alpha. Invest Ophthalmol Vis Sci. 2003;44(11):4885–94.

Osborne N, Block F, Sontag K-H. Reduction of ocular blood flow results in glial fibrillary acidic protein (GFAP) expression in rat retinal Müller cells. Vis Neurosci. 1991;7(06):637–9.

Paris D, Town T, Parker T, Humphrey J, Mullan M. Aβ vasoactivity: an inflammatory reaction. Ann N Y Acad Sci. 2000;903(1):97–109.

Popescu BO, Toescu EC, Popescu LM, Bajenaru O, Muresanu DF, Schultzberg M, Bogdanovic N. 2009 Blood-brain barrier alterations in ageing and dementia. J Neurol Sci. 283(1):99–106.

Prasanna G, Narayan S, Krishnamoorthy RR, Yorio T. Eyeing endothelins: a cellular perspective. Mol Cell Biochem. 2003;253(1–2):71–88.

Privat A, Rataboul P. Fibrous and protoplasmic astrocytes. In: Fedoroff S, Vernadakis A, editors. Astrocytes pt 1: development, morphology, and regional specialization of astrocytes. London: Academic; 1987. pp. 105–29.

Quehenberger P, Bierhaus A, Fasching P, Muellner C, Klevesath M, Hong M, Stier G, Sattler M, Schleicher E, Speiser W. Endothelin 1 transcription is controlled by nuclear factor-kappaB in AGE-stimulated cultured endothelial cells. Diabetes. 2000;49(9):1561–70.

Ramachandran E, Frank RN, Kennedy A. Effects of endothelin on cultured bovine retinal microvascular pericytes. Invest Ophthalmol Vis Sci. 1993;34(3):586–95.

Ren Y, Sun C, Sun Y, Tan H, Wu Y, Cui B, Wu Z. PPAR gamma protects cardiomyocytes against oxidative stress and apoptosis via Bcl-2 upregulation. Vascul Pharmacol. 2009;51(2–3):169–74. doi:10.1016/j.vph.2009.06.004.

Sasaki Y, Takimoto M, Oda K, Früh T, Takai M, Okada T, Hori S. Endothelin evokes efflux of glutamate in cultures of rat astrocytes. J Neurochem. 1997;68(5):2194–200.

Schnitzer J. Retinal astrocytes: their restriction to vascularized parts of the mammalian retina. Neurosci Lett. 1987;78(1):29–34.

Seeram NP. Emerging research supporting the positive effects of berries on human health and disease prevention. J Agric Food Chem. 2011;60(23):5685–6.

Singh R, Kaur N, Kishore L, Gupta GK. Management of diabetic complications: a chemical constituents based approach. J Ethnopharmacol. 2013;150(1):51–70. doi:10.1016/j.jep.2013.08.051.

Song MK, Roufogalis BD, Huang TH. Reversal of the Caspase-dependent apoptotic cytotoxicity pathway by Taurine from *Lycium barbarum* (Goji berry) in human retinal pigment epithelial cells: potential benefit in diabetic retinopathy. Evid Based Complement Altern Med. 2012;2012:323784. doi:10.1155/2012/323784.

Stone J, Dreher Z. Relationship between astrocytes, ganglion cells and vasculature of the retina. J Comp Neurol. 1987;255(1):35–49. doi:10.1002/cne.902550104.

Takeuchi M, Makita Z, Bucala R, Suzuki T, Koike T, Kameda Y. Immunological evidence that non-carboxymethyllysine advanced glycation end-products are produced from short chain sugars and dicarbonyl compounds in vivo. Mol Med. 2000;6(2):114.

Taylor AC, Seltz LM, Yates PA, Peirce SM. Chronic whole-body hypoxia induces intussusceptive angiogenesis and microvascular remodeling in the mouse retina. Microvasc Res. 2010;79(2):93–101.

Tian XM, Wang R, Zhang BK, Wang CL, Guo H, Zhang SJ. Impact of *Lycium Barbarum* polysaccharide and Danshensu on vascular endothelial growth factor in the process of retinal neovascularization of rabbit. Int J Ophthalmol. 2013;6(1):59–61. doi:10.3980/j.issn.2222-3959.2013.01.12.

Verkman AS, Ruiz-Ederra J, Levin MH. Functions of aquaporins in the eye. Prog Ret Eye Res. 2008;27(4):420–33. doi:10.1016/j.preteyeres.2008.04.001.

Wallow I, Burnside B. Actin filaments in retinal pericytes and endothelial cells. Invest Ophthalmol Vis Sci. 1980;19(12):1433–41.

Wang L, Cull GA, Fortune B. Optic nerve head blood flow response to reduced ocular perfusion pressure by alteration of either the blood pressure or intraocular pressure. Curr Eye Res. 2014:1–9. doi:10.3109/02713683.2014.924146.

Xu B, Chibber R, Ruggiero D, Kohner E, Ritter J, Ferro A. Impairment of vascular endothelial nitric oxide synthase activity by advanced glycation end products. FASEB J. 2003;17(10):1289–91.

Yamagishi S-i, Amano S, Inagaki Y, Okamoto T, Koga K, Sasaki N, Yamamoto H, Takeuchi M, Makita Z. Advanced glycation end products-induced apoptosis and overexpression of vascular endothelial growth factor in bovine retinal pericytes. Biochem Biophys Res Commun. 2002a;290(3):973–8.

Yamagishi S, Takeuchi M, Inagaki Y, Nakamura K, Imaizumi T. Role of advanced glycation end products (AGEs) and their receptor (RAGE) in the pathogenesis of diabetic microangiopathy. Int J Clin Pharmacol Res. 2002b;23(4):129–34.

Yang D, Li SY, Yeung CM, Chang RC, So KF, Wong D, Lo AC. *Lycium barbarum* extracts protect the brain from blood-brain barrier disruption and cerebral edema in experimental stroke. PLoS ONE. 2012;7(3):e33596. doi:10.1371/journal.pone.0033596.

Yeh Y-C, Hahm T-S, Sabliov CM, Lo YM. Effects of Chinese wolfberry (*Lycium chinense* P. Mill.) leaf hydrolysates on the growth of *Pediococcus acidilactici*. Bioresour Technol. 2008;99(5):1383–93.

Yeung PKK, Lo ACY, Leung JWC, Chung SSM, Chung SK. Targeted overexpression of endothelin-1 in astrocytes leads to more severe cytotoxic brain edema and higher mortality. J Cereb Blood Flow Metab. 2009;29(12):1891–902.

Yokoi M, Yamagishi S, Takeuchi M, Ohgami K, Okamoto T, Saito W, Muramatsu M, Imaizumi T, Ohno S. Elevations of AGE and vascular endothelial growth factor with decreased total antioxidant status in the vitreous fluid of diabetic patients with retinopathy. Br J Ophthalmol. 2005;89(6):673–5.

Yorio T, Krishnamoorthy R, Prasanna G. Endothelin: is it a contributor to glaucoma pathophysiology? J Glaucoma. 2002;11(3):259–70.

Zhang Y, Stone J. Role of astrocytes in the control of developing retinal vessels. Invest Ophthalmol Vis Sci. 1997;38(9):1653–66.

Zlokovic BV. Neurovascular mechanisms of Alzheimer's neurodegeneration. Trends Neurosci. 2005;28(4):202–8.

Chapter 5
Dermatologic Uses and Effects
of *Lycium Barbarum*

Hui Zhao and Krzysztof Bojanowski

Abstract *Lycium barbarum* (LB) is one of the most intriguing medicinal plants in China. The beauty of its berries combined with the amount of beneficial effects assigned to it would logically make it a strong candidate for skin use, yet relatively few scientific publications address such application. Here, we will review the skin-related effects of oral and topical preparations of LB, based on the published scientific literature and work done in our own laboratory. We will also discuss the obstacles and opportunities for LB in today's dermatological field.

Keywords Wounds · Skin · Aging · Tightening · Wrinkles · Peptidoglycans

5.1 Skin Care

Skin is our largest and most conspicuous organ. It is also the only one, that humans constantly expose to treatments not only to preserve its health, but also its beauty. Historically, these treatments have been botanical in nature. In this respect, our integumentary system is also unique—no other organ in the human body has been target of so many botanical treatment modalities. Hundreds of plant extract combinations have been designed and applied to skin with expectations ranging from the promise of eternal beauty to the cure of syphilis, leper, cancer, and other deadly diseases. The purpose of the vast majority of these preparations is, however, somehow more prosaic—to remedy various forms of inflammatory conditions (eczema, psoriasis…) or to limit water loss by improving the barrier function of the skin.

Lycium barbarum (LB) is well positioned to provide both functionalities. With respect to inflammation, it has been found in our and other laboratories that its fruits (Wang et al. 2002; Chung et al. 2014), leaves (Dong et al. 2009), and bark (Zhang et al. 2013) contain antioxidant components, which provide anti-inflammatory activity, at least partially through the inhibition of the NF-κB signaling (Conner and Grisham 1996; Oh et al. 2012). Animal studies linked this antioxidant activity to a general senescence-inhibitory effect (as measured by a panel of oxidative stress

K. Bojanowski (✉) · H. Zhao
Sunny BioDiscovery, Inc., 972 E. Main St., Santa Paula, CA 93060, USA
e-mail: kbojanowski@sunnybiodiscovery.com

© Springer Science+Business Media Dordrecht 2015
R. C-C. Chang, K-F. So (eds.), *Lycium Barbarum and Human Health*,
DOI 10.1007/978-94-017-9658-3_5

parameters, motor skills, cognition, and nonenzymatic glycation level), bearing further relevance to the antiaging skin care application (Deng et al. 2003; Li et al. 2007; Yi et al. 2013), in agreement with the free-radical theory of aging (Harman 2009).

The predominant LB components with antioxidant activity are peptidoglycans (also called LB polysaccharides, or LBP) (Qiu et al., 2014 and Zhang, 1993); vitamin B, C, taurine, and carotenoids in fruits; while flavonoids, such as rutin prevail in the leaves (Jin et al. 2013). The seeds of Fructus lycii also contain oils, which not only have antioxidant activity, but may improve skin barrier and decrease transepithelial water loss through their ability to interact with the lipid matrix of stratum corneum, such as it is the case of other bioactive oils (Tollesson and Frithz 1993). LB oils may also provide vehicle function facilitating the intracellular absorption of other bioactive components, such as carotenoids (as evidenced by the orange-yellow color). A factor limiting the use of LB oils in cosmetics is the high extraction cost due to low content (unlike peptidoglycans, which represent up to 40 % of the fresh fruit pulp).

Despite the theoretically high potential for beneficial activity, there are only few peer-reviewed studies specifically describing cutaneous benefits of LB. One of them reports that ingesting the aqueous extract (juice) of LB protects mice from UV-induced damage, such as inflammatory oedema, immunosuppression, and sunburn reactions (Reeve et al. 2010). Here also the antioxidant mechanism of action appears to be implicated, as quantified by the inhibition of lipid peroxidation. This study was tangentially corroborated by Wang and collaborators (2011), who reported that pretreatment of human dermal fibroblasts with LBP prior to UVB irradiation saved these cells from G1 growth arrest. Another report comes from our laboratory, where LBP were applied topically on full thickness human skin explants with the result of a selective metalloproteinase (MMP) inhibition. When one of these peptidoglycans (LBGp5) was applied on fibroblasts cultured in suboptimal conditions, it was found to stimulate the production of type I collagen and to promote cell viability (Zhao et al. 2005). Interestingly, LBGp5 is the peptidoglycan with the highest antioxidant activity among the five major LBPs (Huang et al. 2001). This report indicates that the peptidoglycan fraction of LB has a beneficial effect on human skin when applied topically and warrants the development of LBP-based skin care products.

And yet, the Lycium barbarum-containing skin care formulations supported by clinical studies are rare. One reason may be the presence of immunostimulatory components in LB, which may be beneficial for fighting cancer (Tang et al. 2012), but not necessary in skin care. The authors are not aware of any reports of skin reactions to topically applied LB, although allergy to ingested LB has been reported (Monzón Ballarín et al. 2011; Larramendi et al. 2012) in individuals sensitive to multiple food allergens, including tomato (there was cross-reactivity between lipid transfer proteins of LB and tomato, which both belong to the same family—*Solanacea*). Hence the importance of a careful, bioactivity-guided LB fractionation for skin care applications.

Another reason for low skin care use of LB may be its regulatory status in some countries, such as China and Japan, where this herb tends to be perceived as a therapeutic modality and thus is not registered as a cosmetic ingredient. Because it

Fig. 5.1 **a**, **b** Smoothing effect of 3 % Instalift™ Goji solution applied to mature facial skin (**a**: time 0; **b**: time 60 min). Note the transformation in **b** of the initially well-visible (*yellow arrows*) wrinkles in **a**. **c**, **d** Microscopic image of the tightening effect of same solution spread on glass slide (*left* of the meniscus line pointed by *black arrow*), indicating the possible mechanism of action in vivo (mag.: ×40, **c**: time 0; **d**: time 60 min). (Reprinted with permission from RON)

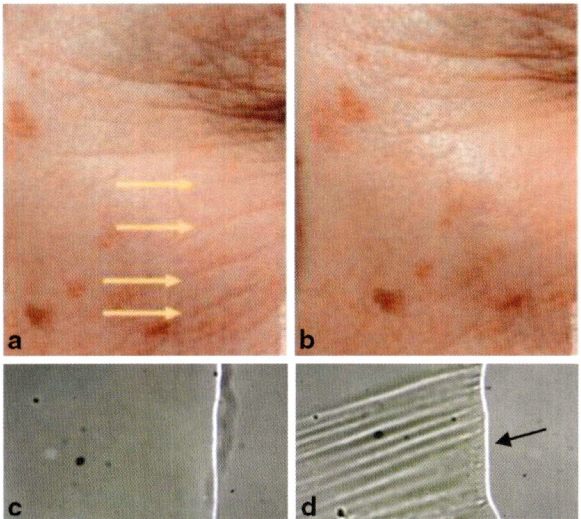

indeed has multiple therapeutic activities, one can say that LB is there a victim of its own success. However, given its widespread dietary and even culinary use, regulatory restrictions specifically aimed at skin care formulations are nevertheless disconcerting. Furthermore, they cannot be substantiated by genotoxic concerns, since LB was demonstrated to be genoprotective rather than genotoxic, using mitomycin C-induced sister chromatid exchange (SCE) in lymphocytes and Ames methods. Same study claimed that SCE in elderly (60 and over) patients ingesting LB polysaccharides was significantly ($p < 0.001$) lower than in the same age control group and became comparable with the SCE rate in young adults (Hong 1995).

In the US, where the topical use of LB is not restricted, two companies—Resources of Nature (RON) and Grant Industries—appear to be leaders in LB ingredient formulations. The clinically tested DC Instalift™ Goji (RON, Fig. 5.1) and Invisaskin GM™ (Grant Industries), which intelligently combine the physicochemical and biological properties of LB peptidoglycans, have over 10 years (as of 2014) of history and underlie many finished skin care products with LB component.

5.2   Wound Healing

Given the above-mentioned beneficial effects of LB on human skin, we searched and failed to locate any publications pertaining to the effect—whether positive or negative—of this medicinal plant on wound healing. Therefore, we conducted a 3-day study on a partial thickness wound model in FT (full thickness) skin substitutes (MatTek, Ashland, MA), where the epidermal layer is peeled off and the underlying dermal layer is exposed. The histochemical Masson trichrome stain at day 3 (Fig. 5.2) shows that compared to the untreated control, the LBP-treated

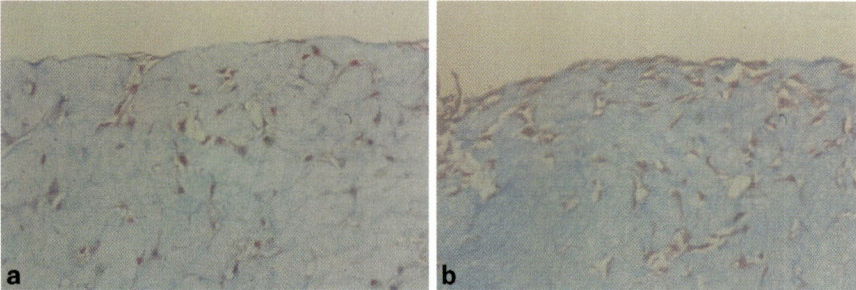

Fig. 5.2 Effect of *Lycium barbarum* polysaccharides (LBP) on early-stage (day 3) wound heal-ing in epidermis-stripped human skin substitutes (MatTek), as visualized by the trichrome stain (mag. ×200). **a** Control (water); **b** LBP (500 µg/mL). Note more intense blue collagen stain and more fibroblasts migrated upward to the wound bed in the LBP-treated tissue as compared with the water control

wounds present a more advanced stage of healing. The difference with controls consists in fibroblasts (dark purple stain), which appear to migrate upward provid-ing topical coverage of the wound, while increasing the collagen output (the blue stain) to facilitate this migration (Sunny BioDiscovery, Inc. unpublished results). This is in agreement with the other two reports of the stimulatory effect of LBP on human dermal fibroblasts (Zhao et al. 2005; Wang et al. 2011) and with the general understanding of the role of the directional migration of fibroblasts during wound healing (Song et al. 2013). However, there is no straight path to LBP-based wound dressings. This is because LB peptidoglycans were reported to potentiate the effect of warfarin—a blood thinner, raising concerns about the adverse effects on blood coagulation (Lam et al. 2001; Ge et al. 2014). Interestingly, both coagulants and an-ticoagulants have been reported to stimulate wound healing (Carney et al. 1992; Fan et al. 2014) and whether LB interferes with this process in the absence of warfarin, as well as the utility of LB in wound dressings remain to be discovered.

Conclusion

LB is a versatile medicinal herb with many health benefits and a few, mostly benign side effects (such as temporary nosebleed following overconsumption (>1 g) of LBP). The dermatologic activities of LB are poorly understood, in part because of the lack of motivation due to the regulatory constrains in Asian countries tradition-ally utilizing in this plant. We hope that this chapter conveys the potential of LB in skin and wound care, and encourages more preclinical and clinical explorations in this direction.

Acknowledgments We would like to thank Stephanie Ma for her expert assistance in this project and George Majewski for helpful discussions.

References

Carney DH, Mann R, Redin WR, et al. Enhancement of incisional wound healing and neovascularization in normal rats by thrombin and synthetic thrombin receptor-activating peptides. J Clin Invest. 1992;89:1469–77.

Chung IM, Ali M, Praveen N, Yu BR, Kim SH, Ahmad A. New polyglucopyranosyl and polyarabinopyranosyl of fatty acid derivatives from the fruits of Lycium chinense and its antioxidant activity. Food Chem. 2014;151:435–43. doi:10.1016/j.foodchem.2013.11.061.

Conner EM, Grisham MB. Inflammation, free radicals, and antioxidants. Nutrition. 1996;12:274–7.

Deng HB, Cui DP, Jiang JM, Feng YC, Cai NS, Li DD. Inhibiting effects of *Achyranthes bidentata* polysaccharide and *Lycium barbarum* polysaccharide on nonenzyme glycation in D-galactose induced mouse aging model. Biomed Environ Sci. 2003;16:267–75.

Dong JZ, Lu da Y, Wang Y. Analysis of flavonoids from leaves of cultivated *Lycium barbarum* L. Plant Foods Hum Nutr. 2009;64:199–204. doi:10.1007/s11130-009-0128-x.

Fan L, Zhou X, Wu P, Xie W, Zheng H, Tan W, Liu S, Li Q. Preparation of carboxymethyl cellulose sulfates and its application as anticoagulant and wound dressing. Int J Biol Macromol. 2014;66:245–53. doi:10.1016/j.ijbiomac.2014.02.040.

Ge B, Zhang Z, Zuo Z. Updates on the clinical evidenced herb-warfarin interactions. Evid Base Complement Altern Med. 2014;66:245–53. doi:10.1016/j.ijbiomac.2014.02.040.

Harman D. Origin and evolution of the free radical theory of aging: a brief personal history, 1954–2009. Biogerontology. 2009;10:773–81. doi:10.1007/s10522-009-9234-2.

Huang LJ, Tian GY, Wang ZF, Dong JB, Wu MP. Studies on the glycoconjugates and glycans from Lycium barbarum L in inhibiting low density lipoprotein (LDL) peroxidation. Acta Pharmaceut Sinica. 2001;36:108–11.

Jin M, Huang Q, Zhao K, Shang P. Biological activities and potential health benefit effects of polysaccharides isolated from Lycium barbarum L. Int J Biol Macromol. 2013;54:16–23. doi:10.1016/j.ijbiomac.2012.11.023.

Lam AY, Elmer GW, Mohutsky MA. Possible interaction between warfarin and *Lycium barbarum* L. Ann Pharmacother. 2001;35:1199–201.

Larramendi CH, García-Abujeta JL, Vicario S, García-Endrino A, López-Matas MA, García-Sedeño MD, Carnés J. Goji berries (*Lycium barbarum*): risk of allergic reactions in individuals with food allergy. J Investig Allergol Clin Immunol. 2012;22:345–50.

Li H. [Study of medicinal effects of goji and its functional ingredients]. Zhongcaoyao. 1995;26,490-5.

Li XM, Ma YL, Liu XJ. Effect of the Lycium barbarum polysaccharides on age-related oxidative stress in aged mice. J Ethnopharmacol. 2007;111:504–11.

Monzón Ballarín S, López-Matas MA, Sáenz Abad D, Pérez-Cinto N, Carnés J. Anaphylaxis associated with the ingestion of goji berries (*Lycium barbarum*). J Investig Allergol Clin Immunol. 2011;21: 567–70.

Oh YC, Cho WK, Im GY, Jeong YH, Hwang YH, Liang C, Ma JY. Anti-inflammatory effect of Lycium fruit water extract in lipopolysaccharide-stimulated RAW 264.7 macrophage cells. Int Immunopharmacol. 2012;13:181–9. doi:10.1016/j.intimp.2012.03.020.

Qiu S, Chen J, Chen X, Fan Q, Zhang C, Wang D, Li X, Chen X, Chen X, Liu C, Gao Z, Li H, Hu Y. Optimization of selenylation conditions for lycium barbarum polysaccharide based on antioxidant activity. Carbohydr Polym. 2014;103:148–53. doi:10.1016/j.carbpol.2013.12.032.

Reeve VE, Allanson M, Arun SJ, Domanski D, Painter N. Mice drinking goji berry juice (*Lycium barbarum*) are protected from UV radiation-induced skin damage via antioxidant pathways. Photochem Photobiol Sci. 2010;9:601–7. doi:10.1039/b9pp00177h.

Song S, Han H, Ko UH, Kim J, Shin JH. Collaborative effects of electric field and fluid shear stress on fibroblast migration. Lab Chip. 2013;13:1602–11. doi:10.1039/c3lc41240g.

Tang WM, Chan E, Kwok CY, Lee YK, Wu JH, Wan CW, Chan RY, Yu PH, Chan SW. A review of the anticancer and immunomodulatory effects of Lycium barbarum fruit. Inflammopharmacology. 2012;20:307–14. doi:10.1007/s10787–011–0107–3.

Tollesson A, Frithz A. Transepidermal water loss and water content in the stratum corneum in infantile seborrhoeic dermatitis. Acta Derm Venereol. 1993;73:18–20.

Wang XY, Wang YG, Wang YF. Ginsenoside Rb1, Rg1 and three extracts of traditional Chinese medicine attenuate ultraviolet B-induced G1 growth arrest in HaCaT cells and dermal fibroblasts involve down-regulating the expression of p16, p21 and p53. Photodermatol Photoimmunol Photomed. 2011;27:203–12. doi:10.1111/j.1600-0781.2011.00601.x.

Wang Y, Zhao H, Sheng X, Gambino PE, Costello B, Bojanowski K. Protective effect of Fructus Lycii polysaccharides against time and hyperthermia-induced damage in cultured seminiferous epithelium. J Ethnopharmacol. 2002;82:169-75.

Yi R, Liu XM, Dong Q. A study of *Lycium barbarum* polysaccharides (LBP) extraction technology and its anti-aging effect. Afr J Tradit Complement Altern Med. 2013;10:171–4.

Zhang X. [Experimental research on the role of *Lycium barbarum* polysaccharide in anti-peroxidation]. Zhongguo Zhong Yao Za Zhi. 1993;18:110–2.

Zhang JX, Guan SH, Feng RH, Wang Y, Wu ZY, Zhang YB, Chen XH, Bi KS, Guo DA. Neolignanamides, lignanamides, and other phenolic compounds from the root bark of Lycium chinense. J Nat Prod. 2013;76:51–8. doi:10.1021/np300655y.

Zhao H, Alexeev A, Chang E, Greenburg G, Bojanowski K. Lycium barbarum glycoconjugates: effect on human skin and cultured dermal fibroblasts. Phytomedicine. 2005;12:131-7.

Chapter 6
Lycium Barbarum and Tumors in the Gastrointestinal Tract

Peifei Li, Bingxiu Xiao, Huilin Chen and Junming Guo

Abstract Chinese wolfberries have been proved to have an antitumor effect and to be less toxic. *Lycium barbarum* polysaccharide (LBP) isolated from aqueous extracts of *L. barbarum* has been identified as one of the major active ingredients that are responsible for the biological activities. LBP significantly affects the vascular endothelial growth factor (VEGF) and cyclin D1 protein expression in liver cancer. It inhibits gastric ulcer development and reduces the incidence of gastric cancer. Induction of cell-cycle arrest participates in the anticancer activity of LBP on gastric cancer cells. LBP has an antitumor effect probably by increasing the number of $CD4^+$ T cells in tumor-infiltrating lymphocytes (TIL) to relieve the immunosuppression and enhance the antitumor function of the immune system. LBP inhibits the growth of colorectal tumor cells by inhibiting the expression of Livin and Bcl-2 and induces the apoptosis of tumor cells. LBP also has a long-term antiproliferative effect on colon cancer. These results provide a theoretical basis for the clinical use of Chinese wolfberries in the prevention and treatment of tumors in the gastrointestinal tract.

Keywords Wolfberry · *Lycium barbarum* · LBP · Hepatocellular carcinoma · Gastric cancer · Esophageal carcinoma · Pancreatic cancer · Colorectal cancer

Tumor, a common type of clinical refractory disease, has become the first death of disease. Tumors cause serious harm to human survival and health. However, conventional strategies for treatment of human tumors are not yet satisfactory. Around half of the drugs currently in clinical use are of natural product origin. Natural products tend to possess well-defined three-dimensional structures, embellished with

J. Guo (✉) · P. Li · B. Xiao
Department of Biochemistry and Molecular Biology,
and Zhejiang Provincial Key Laboratory of Pathophysiology,
Ningbo University School of Medicine,
315211, People's Republic of China,
e-mail: junmingguo@yahoo.com

H. Chen
Ningbo College of Health Sciences, Ningbo 315104,
People's Republic of China

© Springer Science+Business Media Dordrecht 2015   85
R. C-C. Chang, K-F. So (eds.), *Lycium Barbarum and Human Health*,
DOI 10.1007/978-94-017-9658-3_6

functional groups (providing hydrogen bond acceptor/donors, etc.), which have been finest tuned into a precise spatial orientation. For these reasons, natural products possess extraordinary specificity and potency compared to artificially designed molecules. Natural phytochemicals present in medicinal herbs are one of the most attractive approaches in cancer chemotherapy. Studies showed that many Chinese herbal medicines and their active components might have anticancer activities (Gan et al. 2004). The earliest Chinese medicinal monograph documented medicinal use of *Lycium barbarum* around 2300 years ago. The fruit of *Lycium barbarum* of the family *Solanaceae* is well known in traditional Chinese herbal medicine and nowadays has been widely used as a popular functional food (Li 2007). Chinese wolfberries have been proved to have an antitumor effect and to be less toxic. *L. barbarum* polysaccharide (LBP) isolated from aqueous extracts of *L. barbarum* has been identified as one of the major active ingredients that are responsible for the biological activities. These activities include antioxidant properties, antiaging effects, neuroprotection, promotion of endurance, increased metabolism, improved control of glucose and other diabetic symptoms, antiglaucoma effects, immunomodulation, antitumor activity, and cytoprotection (Li 2007; Amagase and Nance 2008).

6.1 LBP and Hepatocellular Carcinoma

6.1.1 Hepatocellular Carcinoma

In all developed regions of the world, hepatocellular carcinoma (HCC) is diagnosed each year in 8.5/105 men and in 3.0/105 women (Ferlay et al. 2004). It has been reported that Asian immigrants or overseas Chinese had higher incidence/mortality rates of liver cancer than people born in their new areas of residence (Grulich et al. 1995; Chen et al. 2002; Tu et al. 2009; Mangtani et al. 2010). HCC is the most common and lethal of all cancers. A diversity of dietary, endogenous, and environmental stimuli mediates hepatocarcinogenesis (Wogan 1989; Guyton and Kensler 1993). At present, the number of HCC patients in China accounted for more than half of the total HCC patients in the world.

6.1.2 The Correlation Between Immune Function of Liver Cancer Patients and Dendritic Cells

Dendritic cells (DCs) can shift to the proliferation of lymphoid organs and stimulate naive T cell activation. As an antigen-specific immune response of the initiator, DCs play an important role in the regulation of immune responses and antitumor processes. In addition, DCs also influence the proliferation of B cells and activate the human immune response.

For the treatment of HCC, one of the most malignant tumors, the commonly used and clinically effective methods include operation, chemotherapy and embolization, low temperature melting, intratumoral injection of drugs, and liver transplantation. However, only a limited number of HCC patients have benefited of these methods. Studies showed that cancer cells secrete immunosuppressive factors. HCC patients show basic cellular immune dysfunction: T cells (CD3$^+$T) and Helper T cells (CD4$^+$T) decrease; and the suppressor T cells (CD8$^+$T) increases; and CD4$^+$/CD8$^+$ significantly lower than normal. The immune function of HCC patients are serious obstacles. In addition, through the release of some antiangiogenic substances (such as IL-12 and interferon γ, INF-γ), DCs may also affect the formation of tumor blood vessels. DCs are potent professional antigen-presenting cells and play an important role in immune responses to start the auxiliary of cytotoxic T cells and T cells (Zhu 2003; Lanzavecchia and Sallusto 2001; Liu 2001). Because of their unique characteristic of immunology, DCs play a key role in the initiation of antitumor immune responses in cancer patients. So, enhancement of DC immune function will enhance the immune function of patients with tumors.

6.1.3   Effect of LBP on Hepatocellular Carcinoma

6.1.3.1   Promotion of Maturation of Dendritic Cells in HCC Patients

LBP has no direct stimulation on the activity of immune cells. However, LBP can combine with DC surface in the presence of polysaccharides and activate DC receptors (Goldman 1988). By these, DCs secrete cytokines and promote T cell activity. At the same time, T cells secrete IL-2, the first signal activator that combines with polysaccharide and has synergistic effect. Studies have shown that LBP could increase the expression of the phenotype of DCs; promote the secretion of IL-12, p70, and IFN-γ; and enhance NF-κB expression (Chen et al. 2012). These results suggest that LBP plays a stronger antitumor role in a virus-related environment. And this phenomenon correlates with the NF-κB-signaling pathway.

6.1.3.2   Induction of Apoptosis

The interaction between the death molecule Fas (APO.1/CD95) and its ligand FasL is one of the important pathways of apoptosis. Fas and its ligand FasL belong to the tumor necrosis factors (TNF) of the transmembrane glycoprotein receptor and the ligand family, respectively. FasL is distributed in activated T cells, natural killer (NK) cells, and mononuclear macrophage immune privilege tissues. The T cell antigen receptor (TCR) in T cells is inducted and then activated. At the same time, the expression of Fas and FasL is induced. Activation of Fas and FasL and their adjacent interaction can be lethal. Activated T cell surface Fas/FasL interactions can directly lead to Dutch act. In addition, FasL can be detached from the cell membrane. As soluble ligands and in an autocrine or paracrine manner, these soluble molecules can act on their host cells and adjacent T cells, respectively. Tumor cells can express FasL, which contacts with activated Fas (+) lymphocytes, and then kill immune cells (Zhu et al. 2005; Abrams 2005). LBP may decrease the expression of FasL in tumor cells and induce immune cells apoptosis. By improving the immune function in the tumor microenvironment, LBP suppresses the growth of tumor cells.

6.1.3.3 Inhibition of the Expression of Some Proteins

Studies have shown that LBP can inhibit the expression of some proteins (Cui et al. 2012). In an *in vivo* study (Cui et al. 2012), rats were divided into four groups. Group A was maintained on basal diet, whereas the remaining three groups (groups B, C, and D) had free access to the basal diet and were orally fed with LBP at 200 mg/kg body weight (b.w.) for group B, 400 mg/kg b.w. for group C, and 600 mg/kg b.w. for group D. Following 4 weeks of this dietary regimen, hepatocarcinogenesis was initiated in all animals by a single intraperitoneal diethylnitrosamine (DENA) injection at a dose of 200 mg/kg b.w. (mixed with peanut oil). Results showed that LBP significantly affected the vascular endothelial growth factor (VEGF) and cyclin D1 protein expression in rat liver cancer.

6.2   LBP and Gastric Cancer

6.2.1   Gastric Cancer

Gastric cancer is a serious health problem and is one of the leading causes of cancer mortality worldwide (Cui et al. 2013). However, conventional strategies for treatment of human gastric cancer are not yet satisfactory.

Malignant gastric tumor originates in the most surface epithelial cells of stomach. Stomach cancer may occur at all sites: in the gastric antrum pyloric area—the largest, in the gastric cardia area—the second, and in the gastric body—slightly less. However, cancer may have different breadth and depth of invasion of the gastric wall. Cancer confined to the mucosa or submucosa is called early gastric cancer. Those that are muscular invasion or metastasis are known as advanced gastric cancer. By naked eye or gastroscopic observation, gastric cancer may be divided into several forms, such as superficial type, mass type, ulcer type, and infiltrative carcinoma (for the cancerization of chronic gastric ulcer). There are many types of cancer cells according to microscopic observation (histological classification), such as adenocarcinoma (accounts for about 90% of cancers, including papillary adenocarcinoma, adenocarcinoma, mucinous adenocarcinoma, and signet-ring cell carcinoma), adenosquamous carcinoma, squamous cell carcinoma, and undifferentiated carcinoma.

6.2.2   Effect of LBP on Gastric Cancer

6.2.2.1   Regulation of Oncogene Expression

The incidence of gastric cancer is a multifactor, multistep, and multistage process of development. Oncogenes, tumor-suppressor genes, apoptosis-related genes, and metastasis-associated genes are involved in these processes. The c-*Myc* gene family belongs to the regulatory gene nuclear proteins. A study showed that the amplification and activation of the c-*Myc* proto-oncogene was the main way of changes in cancer-associated gene expressions. A study showed that PBP can inhibit the expression of c-*Myc* (Hatakeyama et al. 2005).

6.2.2.2   Antibacterial Effects by Regulating the Expression of TNF-α and CGRP

Gastric mucosal dysplasia is an important precancerous gastric lesion. Calcitonin gene-related peptide (CGRP) is the local gastric mucosa defense factor. With vasodilator action, CGRP is transported into the gastric mucosa and gastric cavity, and then protects the gastric mucosa. CGRP is a bioactive peptide composed of 37 amino acids residues. Widely distributed in the gastrointestinal submucosal and myenteric nerve plexus, it has the function of regulating endocrine cells in the gastric antrum. By reducing the cell membrane permeability to calcium ion, CGRP maintains the stability of intracellular calcium ion (Culen et al. 1996). TNF-α produced by endothelial cells regulates mononuclear macrophage activation. With induction of differentiation of T cells and B cells, NK cells enhance killing target cells and promote the phagocytosis of macrophage function. At the same time, TNF-α can increase vascular permeability (Stephens et al. 1988). If TNF-α increases, gastric lesions will inevitably increase (Nishiura et al. 2013). Research has found that the high expression of CGRP in the gastric mucosa of normal rats was associated with low expression of TNF-α. And this situation would induce ulcer in rats. Therefore, the decreased CGRP expression in gastric mucosal tissue may be one of the gastric ulcer-induced factors. TNF-α may be the local inflammatory manifestation of gastric ulcer. Wolfberry leaching liquid has an antibacterial effect. Studies on rats showed that LBP significantly improved the gastric ulcer index by increasing the expression of CGRP and decreasing TNF-α expression in gastric mucosa. It inhibits gastric ulcer development and reduces the incidence of gastric cancer.

6.2.2.3   Inhibiting the Growth of Cancer Cells

The mechanisms of LBP in the growth inhibition of gastric cancer cells have been explored. It is well known that one of the mechanisms of chemotherapeutic drugs on cancer is the interruption of the cell cycle. Our group showed that LBP treatment inhibited growth of gastric cancer MGC-803 and SGC-7901 cells, with cell-cycle arrest in the G_0/G_1 and S phase, respectively (Miao et al. 2010). It is interesting to see that LBP arrested different cell lines from the same types of cancer in different phases. SGC-7901 is a moderately differentiated human gastric adenocarcinoma cell line, and MGC-803 is a poorly differentiated human gastric mucoadenocarcinoma cell line. These results suggest that induction of cell-cycle arrest participates in the anticancer activity of LBP in gastric cancer cells.

Cyclins function as regulators of cyclin-dependent kinase (CDK). Cyclin A binds and activates CDK2 and thus promotes both cell cycle G_1/S and G_2/M transitions. In the G_0/G_1 phase, the main cell cycle regulators are cyclin D, cyclin E, and CDK2. Cyclin E and CDK2 combine together and finally promote cells to go through the G_1–S checkpoint. If the function of these complexes is decreased, cells will be blocked at the G_1–S checkpoint. Cyclin D is not only a cell cycle regulator but is also oncogenic. CDK2 executes a distinct and essential function in the mammalian cell cycle. It is one of the important regulators of the mammalian G_1–S transition. In fact, mutation or inhibition of CDK2 will cause a G_1 block. Mechanisms' study further showed that LBP decreased the level of cyclin D, cyclin E, and CDK2 in MGC-803 cells (Miao et al. 2010). For SGC-7901 cells, LBP increased the level of cyclin A in a dose-dependent manner, while it decreased the expression of CDK2; however, it had no significant effect on the expression of cyclin E. LBP has reliable anticancer activity in human gastric cancer and has potential as a novel, natural anticancer remedy.

6.3 LBP and Esophageal Carcinoma

6.3.1 Esophageal Cancer

Esophageal cancer is one of the most common cancers and a major cause of cancer-related mortality worldwide (Jemal et al. 2010). In 2008, esophageal cancer was with estimated 482,000 new cases the eighth most common malignancy in the world and with 407,000 deaths the sixth most common cause of death (Ferlay et al. 2010). In the USA, the median age of diagnosis of esophageal cancer is 68 years, and about 80 % of all new cases are diagnosed in men. The 5-year overall survival rate increased from 5 % in 1975 to 19% in 2001 (Siegel et al. 2012). In Europe, the mean age of diagnosis is 70 years and 67 % of newly diagnosed cases are males. The 5-year overall survival rate is with 9.8 % much lower than in the USA (Gavin et al. 2012).

The two predominant histological types of esophageal cancer are adenocarcinomas (EAC) and squamous cell carcinomas (ESCC). Both apparently differ in their patterns of incidence and have an own distinct etiology (Holmes and Vaughan 2007). ESCC may be associated with a worse prognosis after surgery than EAC (Siewert et al. 2001).

6.3.2   Effect of LBP on Esophageal Cancer

LBP enhances the immune function and has a significant antitumor effect. It starts from the disease-resistant ability to mobilize the body's own. It promotes the number of nucleated cells and granulocyte/macrophage colonies to increase the number of bone marrow. It can also induce a variety of cytokines, which promote the proliferation, differentiation, and maturation of immune cell activity.

It is believed that LBP, as an effective immune enhancer, has stronger immune activity. It has a very important role in cellular immunity and humoral immunity. Peripheral blood CD4$^+$, CD8$^+$T cell number, and serum cytokine level directly reflect the immune state of the body. Research has found that the number of these cells in peripheral blood of patients with esophageal carcinoma was reduced. LBP injected as a "vaccine" into the intestines, by antigen-presenting cells, causes immunoreceptor tyrosine-based activation motif (ITAM) phosphorylation and activation of T cells (Seder and Ahmed 2003; Wang et al. 2004; Lee et al. 2004; Diederichsen et al. 2003). CD4$^+$T cells are a regulating humoral and cell-mediated immune function of cells. Their differentiation is influenced by many factors, of which the most important factor is the cytokines. IL-2 regulates the immunity in the central position in the network. It promotes peripheral blood mononuclear cells to produce TNF, thereby enhancing the antitumor effect. Because of the immune function and the content of the IL-2 correlation, especially closely related with cell immune function, the IL-2 production is thought as the main index of immune effector cells. NK cells express the IL-2 receptor β chain, a 70kDa moderate affinity receptor, which is directly activated by IL-2. NK cell activation can release IFN-γ and other cytokines in killing tumor cells (Zimmerli et al. 2005). All of these cells respond to IL-2 and obtain nonspecific cytotoxic function. These cells are collectively referred to as the activation of lymphokine killer cells (LAK cells). Compared with LAK cells and NK cells, CD4$^+$T cells not only kill the fresh tumor cells but also have the function of killing many different lineages of tumor cells. Studies show that LBP has an antitumor effect probably by increasing the number of CD4$^+$T cells in tumor-infiltrating lymphocytes (TIL) to relieve the immunosuppression and enhance the antitumor function of the immune system (He et al. 2005).

6.4   LBP and Pancreatic Cancer

6.4.1   Pancreatic Cancer

Pancreatic cancer is the fourth leading cause of cancer death in both men and women in the USA. In 2010, it was estimated that 43,140 Americans will be diagnosed and 36,800 patients will die of pancreatic cancer (Jemal et al. 2010). Approximately, 80 % of pancreatic cancer has metastasized at diagnosis and 5-year survival rate is < 5 % (Jemal et al. 2010). Cigarette smoking, increased body mass index, heavy alcohol consumption, and a diagnosis of diabetes mellitus have been associated with increased pancreatic cancer risk (Lynch et al. 2009; Iodice et al. 2008; Hassan et al. 2007; Michaud et al. 2010; Arslan et al. 2010; Jiao et al. 2010; Everhart and Wright 1995).

6.4.2 Effect of LBP on Pancreatic Cancer

6.4.2.1 Regulation of Oncogenes and Enhanced DNA Repair Function

Pancreatic cancer mainly refers to the pancreatic exocrine gland adenocarcinoma. Molecular biology studies have shown that the activation of oncogenes and inactivation of tumor-suppressor genes and DNA repair genes may play an important role in the development of pancreatic carcinoma. About 90 % of pancreatic cancers have the K-*Ras* gene point mutation at their 12th codon. The genomes of organisms are exposed *in vivo* and in various stress factors, even in very healthy cell integrity of its genomic DNA continued to threaten. Cell damage by periodic repair DNA or respond to DNA damage induced by injury and apoptosis. Cellular responses to DNA damage will result in the abnormal tumor. In the oxidative DNA damage, 8-hydroxyguanine is the highest frequency of product formation, the strongest mutagenicity. It is recognized as an indicator of oxidative stress and oxidative DNA damage. Research shows that LBP regulates oncogenes and enhances DNA repair capacity.

6.4.2.2 The Function of Pancreatic Islet B Cells

Research shows that there is a correlation between new-onset diabetes and pancreatic cancer. Confirmed by experiment, Chinese wolfberries have a significant hypoglycemic effect and have the trend to increase serum insulin. They repair the damaged islet B cells and promote the regeneration of islet B cells' function. This is not all medicines with Chinese wolfberry, so it is hopeful to become a respected novel antidiabetic drug.

Islet-B-cell apoptosis is important in the pathogenesis of type 2 diabetes. Pancreatic duodenal homeohox 1 (PDX-l) is a major regulator of insulin gene transcription. In the regulation of pancreatic development, it plays an important role. It is involved in a series of transcriptional genes in B cells and maintains their characteristics and functions (Mckinnon and Docherty 2001). By upregulating the expression of PDX-1 and insulin, LBP improves the synthesis and secretion of insulin. PDX-1 is an important transcription factor that regulates insulin gene transcription, maturation, and secretion of insulin granules. It uses a combination of insulin gene transcription regulatory region of A3, activates the transcription process, and upregulates the expression of insulin. In addition, unchanged in the transcription level of insulin mRNA, high expression of PDX-1 can improve the intracellular insulin content. Therefore, the PDX-1 determines the expression of the insulin gene promoter and regulates the maturation and secretion of insulin (Alvarsson et al. 2008; Song et al. 2008).

6.5 LBP and Colorectal Cancer

6.5.1 Colorectal Cancer

Colorectal cancer, occurring in any part of the colon, is the most common malignant tumor of digestive tract. In 2009, the estimated number of new cases in the USA was 146,970 (Jemal et al. 2009). In North America and Western Europe, the incidence rate is high. Its incidence is increasing with age; from the age of 40 it begins to rise and reaches its peak at the age of 60–75 years. Colorectal cancer has an obvious geographical distribution. Genetic factors are also involved. Because the tumor grows slowly, early diagnosis is not easy.

6.5.2 Effect of LBP on Colorectal Cancer

6.5.2.1 LBP on Livin and Bcl-2 Expression

Apoptosis is associated with tumor occurrence, development, and resistance (Pan et al. 1997). Apoptosis is regulated by a variety of factors, including the Fas/TNF receptor, Bcl-2 family members, as well as the inhibitor of apoptosis proteins (IAPs). The IAP family is an important regulator of cell apoptosis. It inhibits the apoptosis induced by the cell surface death receptor ligand, cytochrome C, and chemotherapy drugs (Peter et al. 2003). Livin is a new member of the IAP family. It is mainly expressed in some tumor cells, but there is low or no expression in normal tissue (Kasof and Gomes 2001; Lin et al. 2000; Vucic et al. 2000). B cell lymphoma/leukemia-2 (Bcl-2) is a positive anti-apoptosis gene. Bcl-2 was first found in human B cell lymphoma cells. Its gene is located in t(14; 18) chromosomal translocation, a breakpoint. It encodes a 26-kD protein. As an apoptosis inhibition gene, its protein is mainly located in the mitochondrial membrane, nuclear membrane, and endoplasmic reticulum (Tsujimoto and Croce 1986; Wu et al. 2006). Overexpression of Bcl-2 may be related to the occurrence of colorectal cancer. A study found that through inhibiting apoptosis, Bcl-2 extends the life of cells and contributes to tumor occurrence (Sinicrope et al. 1995). It occupies an important position in the development and prognosis of many malignant tumors. Most anticancer drugs play a role by inducing cell apoptosis. Therefore, inducing apoptosis of tumor cells is one of the most common strategies of prevention and treatment of cancer. Studies have shown that the expression of Livin in colorectal cancer tissues was high. However, it is not expressed in normal colorectal mucosa and adenoma. There was a significant correlation between the expression level of Livin and the degree of tumor differentiation

Table 6.1 Inhibitory effect of LBP on tumors in the gastrointestinal tract

Effect	Mechanism	Tumors
Decrease	Oncogene	Gastric cancer, pancreatic cancer
	Expression of several cancer-associated proteins	Liver cancer, gastric cancer, colorectal cancer
	Growth of cancer cells	Gastric cancer, colorectal cancer
Increase	Apoptosis	Liver cancer
	Immune function	Liver cancer, gastric cancer, esophageal cancer, pancreatic cancer, colorectal cancer
	DNA repair capacity	Liver cancer, pancreatic cancer
	Maturation of dendritic cells	Liver cancer

and depth of invasion in colorectal cancer. The Bcl-2 is also expressed higher in colorectal carcinoma tissues compared to normal colorectal mucosa. There was a significant correlation between Bcl-2 expression and depth of tumor invasion. The expression of Livin is positively correlated with the expression of Bcl-2 in colorectal carcinoma. LBP inhibited the growth of colorectal tumor cells by inhibiting the expression of Livin and Bcl-2 and induced the apoptosis of tumor cells.

6.5.2.2 Inhibition of the Growth of Cancer Cells

Our group has showed that LBP inhibited the growth of colon cancer cells (Mao et al. 2011). Human colon cancer SW480 and Caco-2 cells were treated with 100–1000 mg/L LBP for 1–8 days. We found that LBP treatment inhibited colon cancer cell lines in a dose-dependent manner. At concentrations from 400 to 1000 mg/L, LBP significantly inhibited the growth of SW480 cells; while at concentrations from 200 to 1000 mg/L, LBP significantly inhibited the growth of Caco-2 cells. Crystal violet assay showed that LBP had a long-term antiproliferative effect. More importantly, cells were arrested in the G_0/G_1 phase. The changes in cell-cycle-associated protein, cyclins, and CDKs were consistent with the changes in cell-cycle distribution. The results suggest that LBP is a candidate anticancer agent.

In conclusion, the antitumor effect of LBP has been studied extensively. Several mechanisms have been explored (Table 6.1). The results provide a theoretical basis for the clinical use of Chinese wolfberries in cancer prevention and treatment.

Acknowledgments This work was supported by the Student Innovation and Open Laboratory Program of Ningbo University (No. Cxxkf 2008-066), the Student Research and Innovation Program of Ningbo University (2008), the Scientific Innovation Team Project of Ningbo (No. 2011B82014), the Project of Key Disciplines in Ningbo (No. XKL11D2127 and No. XKL11D2128), the Foundation of Zhejiang Provincial Key Laboratory of Pathophysiology (No. 201301), and the K. C. Wong Magna Fund in Ningbo University.

References

Abrams SI. Position and negative consequences of Fas/Fas ligand interactions in the antitumor response. Front Biosci. 2008;10:809–21.

Alvarsson M, et al. Effects of insulin vs. glibenclamide in recently diagnosed patients with type 2 diabetes: a 4-year follow-up. Diabetes Obes Metab. 2008;10(5):421–9.

Amagase H, Nance DM. A randomized, double-blind, placebo controlled clinical study of the general effects of a standardized *Lycium barbarum* (goji) juice, GoChi. J Alt Comp Med. 2008;14(4):403–12.

Arslan AA, et al. Anthropometric measures, body mass index, and pancreatic cancer: a pooled analysis from the pancreatic cancer cohort consortium (PanScan). Arch Intern Med. 2010;170(9):791–802.

Chen CJ, et al. Cancer epidemiology and control in Taiwan: a brief review. Jpn J Clin Oncol. 2002;32 (Suppl):S66–81.

Chen JR, et al. The inducible effect of LBP on maturation of dendritic cells and the related immune signaling pathways in hepatocellular carcinoma (HCC). Curr Drug Deliv. 2012;9(4):414–20.

Cui B, et al. Antitumour activity of Lycium chinensis polysaccharides in liver cancer rats. Int J Biol Macromol. 2012;51(3):314–8.

Cui L; et al. Gastric juice microRNAs as potential biomarkers for the screening of gastric cancer. Cancer. 2013;119(9):1618–26.

Culen JJ, et al. Gastrointestinal myoelectric actively during endotoxemic. Am J Surg. 1996;171(6):596–9.

Diederichsen AC, et al. Prognostic value of the CD4+/CD8+ ratio of the tumour infiltrating Lymphocytes in colorectal cancer and HLA-DR expression on tumour cell. Cancer Immunol Immunother. 2003;52(7):423–8.

Everhart J, Wright D. Diabetes mellitus as a risk factor for pancreatic cancer. A meta-analysis. JAMA. 1995;273(20):1605–9.

Ferlay J, et al. Globocan 2002: cancer incidence, mortality and prevalence worldwide. IARC CancerBase No.5, Version 2.0. Lyon: IARC Press; 2004.

Ferlay J, et al. Estimates of worldwide burden of cancer in 2008: GLOBOCAN 2008. Int J Cancer. 2010;127(12):2893–917.

Gan L, et al. Immunomodulation and antitumor activity by a polysaccharide protein complex from *Lycium barbarum*. Int Immunopharmacol. 2004;4(4):563–9.

Gavin AT, et al. Oesophageal cancer survival in Europe: a EUROCARE-4 study. Cancer Epidemiol. 2012;36(6):505–12.

Goldman R. Characteristics of the beta-glucan receptor of murine macrophages. Exp Cell Res. 1988;174(2):481–90.

Grulich AE, et al. Cancer incidence in Asian migrants to New South Wales, Australia. Br J Cancer. 1995;71(2):400–8.

Guyton KZ, Kensler TW. Oxidative mechanisms in carcinogenesis. Br Med Bull. 1993;49(3):523–44.

Hassan MM, et al. Risk factors for pancreatic cancer: case-control study. Am J Gastroenterol. 2007;102(12):2696–707.

Hatakeyama S, et al. Targeted destruction of c-Myc by an engineered ubiquitin ligase suppresses cell transformation and tumor formation. Cancer Res. 2005;65(17):7874–9.

He YL, et al. Effects of *Lycium barbarum* polysaccharide on tumor microenvironment T-lymphocyte subsets and dendritic cells in H22-bearing mice. Zhong Xi Yi Jie He Xue Bao. 2005;3(5):374–7.

Holmes RS, Vaughan TL. Epidemiology and pathogenesis of esophageal cancer. Semin Radiat Oncol. 2007;17(1):2–9.

Iodice S, et al. Tobacco and the risk of pancreatic cancer: a review and meta-analysis. Langenbecks Arch Surg. 2008;393(4):535–45.

Jemal A, et al. Cancer statistics, 2009. CA Cancer J Clin. 2009;59(4):225–49.

Jemal A, et al. Cancer statistics, 2010. CA Cancer J Clin. 2010;60(5):277–300.

Jiao L, et al. Body mass index, effect modifiers, and risk of pancreatic cancer: a pooled study of seven prospective cohorts. Cancer Causes Control. 2010;21(8):1305–14.

Kasof GM, Gomes BC. Livin, a novel inhibitor of apoptosis protein family member. J Biol Chem. 2001;276(5):3238–46.

Lanzavecchia A, Sallusto F. Regulation of T cell immunity by dendritic cells. Cell. 2001;106(3):263–6.

Lee BN, et al. Depressed type 1 cytokine synthesis by superantigen-related high-grade squamous intraepithelial lesions. Clin Diag Lab Immunol. 2004;11(2):239–44.

Li XM. Protective effect of *Lycium barbarum* polysaccharides on streptozotocin-induced oxidative stress in rats. Int J Biol Macromol. 2007;40(5):461–5.

Lin JI-I, et al. KIAP, a novel member of the inhibitor of apoptosis protein family. Biochem Biophys Res Commun. 2000;279(3):820–31.

Liu YJ. Dendritic cell subsets and lineages, and their functions in innate and adaptive immunity. Cell. 2001;106(3):259–62.

Lynch SM, et al. Cigarette smoking and pancreatic cancer: a pooled analysis from the pancreatic cancer cohort consortium. Am J Epidemiol. 2009;170(4):403–13.

Mangtani P, et al. Cancer mortality in ethnic South Asian migrants in England and Wales (1993–2003): patterns in the overall population and in first and subsequent generations. Br J Cancer. 2010;102(9):1438–43.

Mao F, et al. Anticancer effect of *Lycium barbarum* polysaccharides on colon cancer cells involves G0/G1 phase arrest. Med Oncol. 2011;28(1):121–6.

Mckinnon CM, Docherty K. Pancreatic duodenal homeobox-1, PDX-1, a major regulator of beta cell identity and function. Diabetologia. 2001;44(10):1203–14.

Miao Y, et al. Growth inhibition and cell-cycle arrest of human gastric cancer cells by *Lycium barbarum* polysaccharide. Med Oncol. 2010;27(3):785–90.

Michaud DS, et al. Alcohol intake and pancreatic cancer: a pooled analysis from the pancreatic cancer cohort consortium (PanScan). Cancer Causes Control. 2010;21(8):1213–25.

Nishiura H, et al. Interleukin-21 and tumor necrosis factor-α are critical for the development of autoimmune gastritis in mice. J Gastroenterol Hepatol. 2013;28(6):982–91.

Pan H, et al. Apoptosis and cancer mechanism. Cancer Surv. 1997;29:305–27.

Peter L, et al. The inhibitors of apoptosis: there is more to life than Bcl-2. Oncogene. 2003;22(53):8568–80.

Seder RA, Ahmed R. Similarities and differences in CD4+ and CD8+ effector and memory T cells generation. Nat Immunol. 2003;4(9):835–42.

Siegel R, et al. Cancer treatment and survivorship statistics, 2012. CA Cancer J Clin. 2012;62(4):220–41.

Siewert JR, et al. Histologic tumor type is an independent prognostic parameter in esophageal cancer: lessons from more than 1000 consecutive resections at a single center in the Western world. Ann Surg. 2001;234(3):360–7.

Sinicrope FA, et al. Bcl-2 and P53 oncoprotein expression during colorectal tumorigenesis. Cancer Res. 1995;55(2):237–41.

Song B, et al. Chop deletion reduces oxidative stress, improves B cell function and promotes cell survival in multiple mouse models of diabetes. J Clin Invest. 2008;118(10):3378–89.

Stephens KE, et al. Tumor necrosis factor cause increased pulmonary and edema. Comparison to septic acute lung injury. Am Rev Respir Dis. 1988;137(6):1364–70.

Tsujimoto Y, Croce CM. Analysis of the structure, transcripts, and protein products ofbcl-2, the gene involved in human follicular lymphoma. Proc Natl Acad Sci U S A. 1986;83(14):5214–8.

Tu RSP, et al. A cross-border comparison of hepatitis B testing among Chinese residing in Canada and the United States. Asian Pac J Cancer Prev. 2009;10(3):483–90.

Vucic D, et al. ML-IAP, a novel inhibitor of apoptosis that is preferentially expressed in human melanomas. Curr Biol. 2000;10(21):1359–66.

Wang R, et al. Induction in humans of CD8+ and CD4+ T cell and antibody responses by sequential immunization with malaria DNA and recombinant protein. J Immunol. 2004;172(9):5561–9.

Wogan GN. Dietary risk factors for primary hepatocellular carcinoma. Cancer Detect Prev. 1989;14(2):209–13.

Wu J, et al. Significance of apoptosis and apoptofic-related proteins, Bcl-2 and Bax in primary breast cancer. Breast J. 2006;6(1):44–52.

Zhu KJ. Dendritic cell and autoimmune diseases. Zhejiang Da Xue Xue Bao Yi Xue Ban. 2003;32(1):81–4.

Zhu Q, et al. Mechanism of counter attack of colorectal cancer cell by Fas/Fas ligand system. World J Gastroenterol. 2005;11(39):6125–9.

Zimmerli SC, et al. HIV-1-specific INF-γ/IL-2-secreting CD8 T cells support CD4-independent proliferation of HIV-1-specific CD8 T cells. Proc Natl Acad Sci U S A. 2005;102(20):7239–44.

Chapter 7
Prevention of Neurodegeneration for Alzheimer's Disease by *Lycium barbarum*

Yuen-Shan Ho, Xiao-ang Li, Clara Hiu-Ling Hung and Raymond Chuen-Chung Chang

Abstract Alzheimer's disease (AD) is a chronic neurodegenerative disease that will affect more than 50 million people worldwide. AD is an aging-associated disease because the chance of suffering from AD is double after age of 65. This is a complex disease that is highly related to our environmental stress, experience of head trauma, our daily diet and exercise, quality of our sleep, and even air pollution. All these factors give multiple hits to our brain promoting neurodegeneration in the aging processes. As *Lycium barbarum* (Wolfberry) is an antiaging Chinese medicine, we have attempted to understand the scientific meaning of this antiaging medicine. In this chapter, we will at first summarize different risk factors leading to AD. Then, we will discuss how Wolfberry protects neurons against AD. Furthermore, we will discuss how Wolfberry prevents those risk factors leading to AD. As Chinese medicine emphasizes on prevention of disease by adjusting the whole body

R.C.C. Chang (✉) · C.H.L. Hung
Laboratory of Neurodegenerative Diseases, Department of Anatomy, LKS Faculty of Medicine, The University of Hong Kong, Pokfulam, Hong Kong SAR, People's Republic of China

Y.-S. Ho (✉)
School of Nursing, Faculty of Health and Social Sciences, The Hong Kong Polytechnic University, Hung Hom, Kowloon, Hong Kong SAR, People's Republic of China
e-mail: janice.ys.ho@polyu.edu.hk

X.-a. Li
State Key Laboratory of Quality Research in Chinese Medicine, Macau University of Science and Technology, Macau SAR, People's Republic of China

C.H.L. Hung
Institute of Chinese Medicinal Science, University of Macau, Macau SAR, People's Republic of China

R.C.C. Chang
Research centre of Heart, Brain, Hormone and Healthy Aging, LKS Faculty of Medicine, The University of Hong Kong, Pokfulam, Hong Kong SAR, People's Republic of China
e-mail: rccchang@hku.hk

State Key Laboratory of Brain and Cognitive Sciences, The University of Hong Kong, Pokfulam, Hong Kong SAR, People's Republic of China

© Springer Science+Business Media Dordrecht 2015
R. C-C. Chang, K-F. So (eds.), *Lycium Barbarum and Human Health*,
DOI 10.1007/978-94-017-9658-3_7

health from a holistic point of view, our understanding of Wolfberry will help us to advance our knowledge of how this natural product elicits multi-target effects to prevent the devastating problem of dementia.

Keywords Alzheimer's disease · Aging · Depression · Diabetes mellitus · Hyperhomocysteinemia · β-Amyloid peptide · Glutamate toxicity

7.1   Introduction

Alzheimer's disease (AD) is the major cause of dementia. As the average lifespan increases, AD has already become the fourth leading cause of death in high-income countries. AD has drawn great attention because it brings great impact on the society. In the USA, the estimated aggregate payment for health care, long-term care, and hospice for people with AD and other dementia was US$ 200 billion in 2012, and it is expected to increase to US$ 1.1 trillion in 2050. For each health caregiver (family members or other unpaid caregivers), an average of 21.9 h/week is devoted to the dementia patients. The economic effect of dementia will be greater than that of cancer, heart disease, and stroke combined due to the aging population (Alzheimer's disease association 2012). Therefore, there is an urgent need for the development of promising therapeutic means for the treatment and prevention of AD. Today, the major drugs used are cholinesterase inhibitors and memantine, which provide modest therapeutic effect to relieve the symptoms but yet do not cure the disease. Since the cause of AD is still unknown, and there are many failures of disease-modifying therapies in different steps of clinical trials, more experts suggest that we should target the risk factors of AD rather than just focusing on the pathological hallmarks, i.e., senile plaques and neurofibrillary tangles (Korczyn 2012). In this regard, the herbal medicine emphasizing a holistic approach in treating diseases seems to match with this concept. Wolfberry (the fruit of *Lycium barbarum*) is a common functional food and herb in Asian countries such as China, Korea, and Japan. It has been considered to be an antiaging herb; and therefore, it has been widely used for the treatment of aging-associated diseases (Ho et al. 2011). In the recent decades, a number of studies showed that Wolfberry had neuroprotective properties and it could reduce neurodegeneration related to AD. More importantly, Wolfberry has been shown to attenuate neuronal damages induced by AD risk factors. Since delaying both disease onset and progression by as little as 1 year could potentially lower the prevalence of AD by 9.2 million in 2050 (Brookmeyer et al. 2007), any therapy or drug that has a preventive role can bring significant impact on the disease and health care system. In this chapter, we will at first discuss the multiple factors leading to aging and degenerative changes in AD. Then, we will further review how Wolfberry protects neurons against neurodegeneration in AD through antagonizing multiple risk factors.

7.2 Multiple Hits on the Brain—Risk Factors of AD and the Actions of Wolfberry

The most prominent pathological hallmarks of AD are senile plaque and neurofi-brillary tangle. The former is mainly made up of β-amyloid peptide (Aβ) and the lat-ter is mainly made up of a hyperphosphorylated form of the microtubule-associated tau protein. Both Aβ and hyperphosphorylated tau can impose cytotoxicity and im-pair the normal function of neurons; and hence researchers believed that they can be the therapeutic targets of AD for decades (Querfurth and LaFerla 2010). Although the progressive clinical deterioration in AD is accompanied by plaques and tangles, no solid evidence proved that these histological changes are pathologically causal to the cognitive decline in human. A number of upstream events and processes or a combination of them can result in cognitive impairment and finally lead to the formation of plaques and tangles. Epidemiological studies have identified many risk factors for developing AD and some of them are modifiable. These risk factors include environmental toxin, lifestyle, and other health conditions. Half of the AD cases worldwide are probably attributed to these modifiable risk factors (Barnes and Yaffe 2011). Therefore, concerted aggressive action against these risk factors is important to fight AD (Korczyn 2012).

7.2.1 Age

Age is known as the strongest risk factor being identified so far for AD. Only 1 % of people aged 60–65 years old have AD, but the prevalence increases to >30 % in people aged 85 years old. The prevalence of AD doubles every 5 years after the age of 60 (Hebert et al. 2001). It is not fully understood what age-related changes af-fect the neurodegeneration process. One major proposed mechanism is the increas-ing levels of oxidative stress during aging (Harman 2006; Holliday 2006). Most free radicals are produced in the mitochondria. When we become old, mitochondria produce more free radicals for the same amount of adenosine triphosphate (ATP) they made. In the same time, the lipid membrane begins to leak and therefore free radicals escape from the mitochondria to the cytosol. The defense mechanisms such as the production of antioxidative enzymes reduce. All these factors lead to high levels of oxidative stress (Anton et al. 2005). Markers of oxidative stress in cere-brospinal fluid (CSF) of AD patients are higher than that in the control subjects and they were negatively correlated with the MMSE (Mini Mental State Examination) score (Ahmed et al. 2005). Experiment has also demonstrated that increased oxida-tive stress promotes production of Aβ through modulating the activity of γ-secretase (Shen et al. 2008). Targeting oxidative stress can therefore be one of the potential strategies to prevent or delay AD.

Wolfberry has demonstrated beneficial effects in aging animal models. Firstly, oral feeding of *L. barbarum* polysaccharide (LBP) for 30 days markedly increased the levels of superoxide dismutase (SOD), catalase (CAT), glutathione peroxidase

(GSH-Px), and total antioxidant capacity in the lung, liver, heart, brain, and serum of aged mice. The level of malondialdehyde (MDA) in these mice was decreased. LBP also reversed age-induced reduction of thymus and spleen weight and improved the function of macrophage. These data suggest that LBP from Wolfberry promote the antioxidant capacity and immune function in aged mice (Li et al. 2007). In the second study, aging in mice was induced by injection of D-galactose. Oral administration of LBP significantly reduced D-galactose-induced elevation of lipid peroxidation, lipofuscin, and monoamine oxidase B (MAO-B) in the brains, and improved their cognitive functions (Tang and He 2013). In fact, the antioxidative effect of Wolfberry has also been documented in various studies (Cheng and Kong 2011; Cui et al. 2011). Collectively, these findings suggest that Wolfberry has the potential to antagonize high levels of oxidative stress during aging, at least in animals, and therefore might be able to reduce aging-associated degeneration (Fig. 7.1).

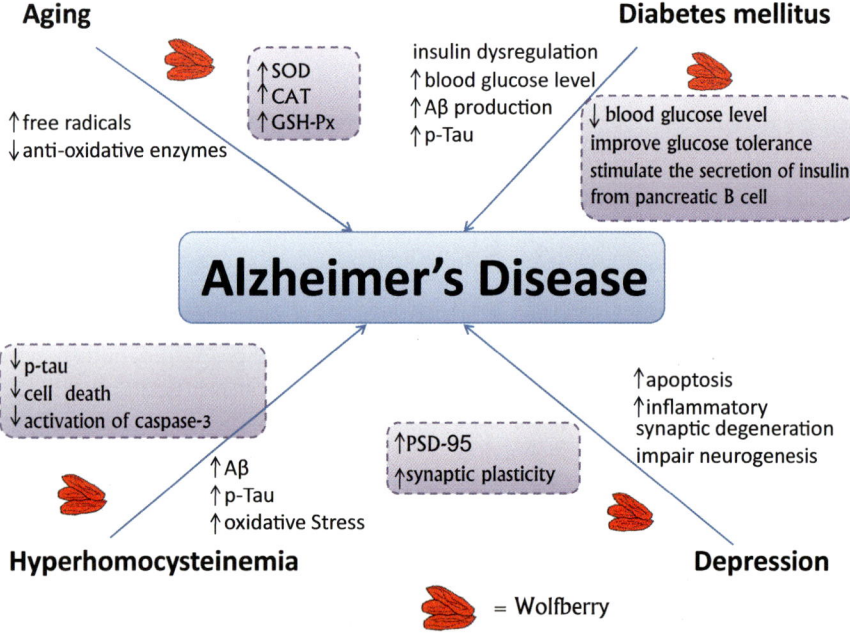

Fig. 7.1 Potential effects of Wolfberry on the prevention of AD. The development of AD is affected by various risk factors, which can increase the level of oxidative stress, neuroinflammation and induce AD-related pathological changes. Wolfberry can antagonize some of these changes hence may play a protective role for AD. *SOD* superoxide dismutase, *CAT* catalase, *GSH-Px* glutathione peroxidase, *Ab* β-amyloid peptide

7.2.2 Diabetes Mellitus

Diabetes has been associated with an increased risk of AD and dementia. Several prospective, population-based studies reported an increased risk of AD in patients with diabetes (Kopf and Frolich 2009). While the worldwide prevalence of diabetes is about 6.4 %, it has been estimated that a 10 % reduction of the prevalence of diabetes would result in 81,000 less AD cases (Barnes and Yaffe 2011). Animal studies have shown that impair glucose metabolism can affect synaptic plasticity, causing deficits in long-term potentiation and long-term depression (Artola et al. 2005; Kamal et al. 2006). A number of possible mechanisms explain how diabetes can promote degeneration in the brain. The most well discussed one is insulin resistance. In AD patients, the levels of insulin were found to decrease in the CSF and increase in plasma. The observed changes were greater for patients with advanced AD, suggesting a possible correlation between insulin and disease severity (Craft et al. 1998). Another piece of supporting evidence is that AD patients and normal subjects react to insulin differentially. When insulin is administrated, the levels of plasma Aβ reduced in the normal subjects while that increased in the AD patients (Kulstad et al. 2006). On the other hand, animal models of AD showed that increased productions of Aβ promote insulin dysregualtion. Amyloid precursor protein (APP)/presenilin 2 (PS2) coexpression in diabetes-positive (db/+) mice results in hyperinsulinemia, hyperglycemia, and hypercholesterolemia together with augmented β-cell mass in the pancreas (Jimenez-Palomares et al. 2012). These data suggest a vicious cycle form between Aβ production and insulin dysregulation. The intracellular signaling cascade of insulin also affects the phosphorylation of tau. Insulin is known to activate phosphatidylinositol 3-kinase (PI3K), which phosphorylates and activates protein kinase B (Akt). Activated Akt can inactivate glycogen synthase kinase-3β, which is a major kinase to phosphorylate tau in the brain. Besides, poor glycemic control in diabetes increases the levels of advanced glycation products (AGE) and oxidative stress in the brain, which can promote neurodegeneration in the central nervous system (CNS; Wrighten et al. 2009).

The hypoglycemic effect of Wolfberry has been documented in several studies (Jing et al. 2009; Luo et al. 2004; Zhu et al. 2013). In the study conducted by Luo and colleagues, feeding of LBP reduced blood glucose level and serum total cholesterol and triglyceride levels in alloxan-induced diabetic rats. They reported that the blood glucose levels decreased in all alloxan-treated rabbits fed with Wolfberry water extract, crude LBP and LBP-X (a fraction of total LBP); and these animals had improved glucose tolerance (Luo et al. 2004). Similar findings were also reported by Jing and colleagues, who found that 28 days LBP treatment significantly reduced the fasting blood glucose levels, total cholesterol, and triglyceride levels in diabetic mice (Jing et al. 2009). The hypoglycemic effect of LBP was further confirmed in a high-fat diet + streptozotocin mice diabetes model (Zhu et al. 2013). In the same study, researchers found that LBP-enhanced glucose consumption in HepG2 cells and 3TE-L1 cells, and stimulated the secretion of insulin from pancreatic β cell. The dual actions of LBP on the pancreatic cells and peripheral tissue may explain the hypoglycemic action of LBP. In a pilot study, healthy human subjects received Wolfberry juice found to have increased caloric expenditure after a single bolus of the juice intake and a reduction of waist circumference after 14 days consumption (Amagase and Nance 2011). All these data support that Wolfberry is likely to have beneficial effects on diabetes patients and thus be potential to reduce the risk of neurodegeneration in AD.

7.2.3 Depression

Neuropsychiatric symptoms are often observed throughout the course of AD and they may severely impair quality of life. Depression is frequently found in AD patients. Meta-analysis found that people with depression had about a two times increased risk of dementia and AD compared with those without a history of the disease (Jorm 2001; Ownby et al. 2006). Anxiety and depression have been shown to accelerate cognitive decline in AD patients (Starkstein et al. 2008). However, the use of traditional antidepressant in elderly with depression does not always results in improved cognitive functions (Butters et al. 2000; Knegtering et al. 1994). Therefore, it seems that prevention of depression is important for reducing the risk of cognitive impairment. Moreover, it has been estimated that more than 10 % of AD (nearly 3.6 million) worldwide are potentially attributable to depression, which further support the important role of depression in AD (Barnes and Yaffe 2011). Several pathological and biological linkages of depression and AD have been proposed (Wong et al. 2013; Wuwongse et al. 2010, 2013). For instance, in animal models of depression and postmortem brains of depressed subjects, decrease in anti-apoptotic markers was observed. This can make neurons more vulnerable to the degenerative factors in AD. The levels of pro-inflammatory cytokines also increased in depression subjects. These pro-inflammatory cytokines influence neuronal functions in the prefrontal cortex and hippocampus by modulating the transmission of neurotransmitters and growth factors (Leonard and Myint 2006; Poon et al. 2013).

Wolfberry has demonstrated antidepressant effect in a depression animal model conducted by Zhang and colleagues (Zhang et al. 2012). They first fed the rats with LBP for 7 days and then corticosterone was injected subcutaneously for 14 days to induce depressive-like behavior. The group found that corticosterone treatment significantly increased the rats' immobility period during force swimming test, which is widely used for detecting depressive symptoms in rodents. Rats fed with LBP had significantly shorter immobility period, suggesting lesser degree of depressive-like symptoms. Furthermore, LBP restored the impairment of neurogenesis in the hippocampus of corticosterone-injected rats, although it was found that it is not directly related to the observed behavioral changes. LBP also restored spine density and the decreased level of PSD-95 (a postsynaptic protein which is important in neuronal plasticity) in the hippocampus, suggesting that LBP enhanced synaptic plasticity in the region. Data from this animal study support the notion that Wolfberry has a preventive or even a therapeutic role in depression. Further studies are required to elucidate the exact mechanism or to confirm the effects of LBP in different depression animal models. Recently, a US health product company conducted some studies on the effects of their Wolfberry juice. Although larger scale of experiment is required before definite conclusion is drawn, their pilot findings demonstrated that human subjects who drink Wolfberry juice had additional improvements in fatigue, depression, and sleeping quality (Paul Hsu et al. 2012). It is hoped that more in-depth studies can be conducted to verify the effects of Wolfberry on human.

7.2.4   Hyperhomocysteinemia

Homocysteine is a nonessential amino acid whose metabolism depends on folate, vitamin B6, and vitamin B12. Elevated plasma levels of total homocysteine are associated with increased risk for dementia and AD (McIlroy et al. 2002; Morris 2003). It is one of the nongenetic risk factors that scientist are interested to target for the prevention of AD. A randomized controlled study showed that high homocysteine levels at baseline are associated with faster gray matter atrophy in elderly subjects. Supplement of vitamin B as treatment can reduce homocysteine level and attenuate the cerebral atrophy in those grey matter regions specifically vulnerable to the AD process by as much as sevenfold (Douaud et al. 2013). Autopsy findings also confirm the pathological linkage between homocysteine level and AD. In a population aged ≥ 85 years, elevated baseline homocysteine has been shown to associate with increased neurofibrillary tangles in their postmortem brains. There is also an association between Aβ accumulation and elevated homocysteine. In the same time, those with homocysteine levels in the highest quartile also had more severe medial temporal atrophy and periventricular and deep white matter hyperintensities, which is associated with increased risk of dementia (Hooshmand et al. 2013; Debette and Markus 2010). Together with several in vitro and in vivo studies, it is believed that high levels of homocysteine promote the production of Aβ and the generation of oxidative stress, enhance tau phosphorylation, and even affects normal cell cycle states (Morris 2003).

Data from our laboratory showed that Wolfberry could antagonize homocysteine neurotoxicity in vitro (Ho et al. 2010). Primary cultures of cortical neurons are susceptible to the toxicity of homocysteine, which can lead to apoptosis and cell death. LBP significantly attenuated homocysteine-induced neuronal cell death and the activation of caspase-3. LBP also reduced homocysteine-induced tau pathology. It is known that hyperphosphorylation of tau can affect normal neuronal functions. With LBP treatment, the increased levels of tau phosphorylation at Ser198/199/202, Ser396, and Ser214 by homocysteine were marked reduced. Moreover, the level of toxic truncated-tau was also decreased accordingly. It was found that LBP could reduce the phosphorylation of extracellular-signal-regulated kinase (ERK) and c-Jun-N-terminal kinase (JNK) but not the phosphorylation of glycogen synthase kinase 3 beta (GSK3β). These changes in tau kinases might partly explain the protective effects of LBP. Research from our group also proved that tau phosphorylation is highly regulated by endoplasmic reticulum stress (ER stress) during neurodegeneration and blocking of ER stress could reduce tau hyperhosphorylation (Ho et al. 2012). Our previous findings showed that LBP could reduce dithiothreitol (DTT)-induced neurotoxicity (Yu et al. 2006). We, therefore, further investigate if LBP can also reduce homocysteine-induced phosphorylation of tau through modulating the ER stress signaling.

7.3 Direct Protective Action of Wolfberry on the AD Brain

7.3.1 Aβ Neurotoxicity

Aβ is the major component found in senile plaques and its CSF levels has been considered as a biomarker for AD. Numerous studies prove that Aβ is toxic to neurons (please refer to Cavallucci et al. 2012 for a detailed review). Dysfunction of synapse is found in all stages of AD. The amount of presynaptic vesicles protein is reduced throughout the disease progression. It was found that Aβ accumulated at synapses and induces a progressive toxicity. It increases N-methyl-D-aspartate receptor (NMDAR)-mediated Ca^{2+} influx, thus disrupts neuronal transmission. This can lead to subsequent pathological events such as the generation of reactive oxygen species. It also impairs the normal function of mitochondria and triggers the activation of caspase-3 and apoptosis. The activation of capsase-3 can impair NMDAR-dependent long-term depression.

Our laboratory found that LBP could attenuate the activation of caspase-3 and neuronal cell death induced by $Aβ_{1-42}$ and $Aβ_{25-35}$, which is a shorter synthetic fragment. Interestingly, this protective effect was only provided by the polysaccharide fraction of Wolfberry but not other fractions including betaine and β-sitosterol (Ho et al. 2007, Yu et al. 2005). We also compared the protective effects of LBP with a well-known Western neuroprotective medicine, lithium chloride (LiCl), and discovered that LBP even has a wider effective and safety dosage (Yu et al. 2005, 2007). Our data supported that different fractions of LBP protected neurons through distinct mechanisms. The water extract of Wolfberry contained LBP fractions that could inhibit some proapoptotic signaling pathways, including the JNK and double-stranded RNA-dependent protein kinase (PKR; Yu et al. 2005, 2007); while the alkaline extract contains LBP that could upregulate the pro-survival Akt pathway (Ho et al. 2007). In fact, Wolfberry was found to promote the expression of genes related to cell survival in mouse spleen. In mice that received 7 days of oral feeding of micronized Wolfberry, the gene expression of tumor necrosis factor (TNF), NF-kappaB, and Bcl-2 were upregulated, while that of APAF-1 and caspase-3 were downregulated (Lin et al. 2011). This indirect evidence suggested that Wolfberry could affect multiple gene expression and these changes might explain how it helps our body to cope with stress and reduces damages.

7.3.2 Glutamate Neurotoxicity

Glutamatergic NMDA receptors are important for learning and memory. However, overactivation of the NMDA receptors can lead to Ca^{2+} overload and excitotoxicity, which disturb neuroplasticity and cause neuronal cell death. Glutamate excitotoxicity is involved in the pathogenesis of AD. Apart from the excitotoxicity, high concentrations of glutamate can induce oxidative stress by inhibiting the uptake of cysteine thus the synthesis of glutathione (Murphy et al. 1990). Therefore, reducing glutamate excitotoxicity or restoring the homeostatsis in the glutamatergic system is one of the available strategies for the treatment of AD. Currently, the noncompetitive NMDA receptor antagonist memantine has been approved by the European Medicines Agency (EMEA) and Food and Drug Administration (FDA) for the treatment of moderate-to-severe AD (Ho et al. 2009; Hynd et al. 2004).

It has been reported that LBP from Wolfberry antagonized the toxicity of glutamate in primary cultures of cortical neurons (Ho et al. 2009). Pretreatment of neurons with LBP significantly reduced the necrosis and apoptosis following subsequent glutamate exposure. The protective effects of LBP were even comparable to that provided by memantine, which further support the potential of LBP as neuroprotective drug for AD. Interestingly, the LBP used in this study had no antioxidant effects and it could not reduce the levels of reactive oxygen species triggered by hydrogen peroxide. It was suggested that LBP might mediate its protection by attenuating the phosphorylation of the apoptotic kinase JNK. Since glutamate toxicity is also involved in other diseases such as glaucoma and stroke, it is worth to conduct further studies on Wolfberry.

7.4 Wolfberry Preserves Memory Functions in Various Animal Models

Cognitive impairment remains the major clinical symptoms in AD. Cognitive performance is always used as the primary and almost the most important parameter for the diagnosis of AD. It is, therefore, important to evaluate the efficacy of potential drugs by observing if there are cognitive improvements after intake. Two animal studies proved that Wolfberry could improve cognition (Chen et al. 2014; Zhang et al. 2013).

In the first study, Wolfberry water extract was fed to APP/PS double transgenic mice for 2 weeks, and then their cognitive performance was measured by Morris water maze test. In this test, mice were put in a water tank and they were trained to find a hidden platform with visual cues. The time they spent in searching the hidden platform during the training period and the final stage (probe trial) indicated their learning and memory functions. APP/PS double transgenic mice are used commonly as animal model for AD. These mice have spontaneous increase in Aβ production and memory loss when they grow up. The researchers found that mice fed with Wolfberry water extract took shorter time to find the hidden platform and they spent longer time in the memorized zone during the probe trial when compared to those just fed with saline, indicating these APP/PS1 mice had better learning and memory. ELISA experiment confirmed that the Wolfberry water extract reduced the levels of $A\beta_{1-42}$ in these mice, which explained the observed beneficial effects (Zhang et al. 2013).

In the second study, rats received chronic infusion of scopolamine (SOC) to induce memory impairment. Scopolamine can inhibit the muscarinic acetylcholine receptor and leads to characteristic cognitive and memory deficits of AD. In the group pretreated with LBP, the rats had better performance in Morris water maze test as well as in novel object and object location recognition tasks, indicating an improvement on memory. It was reported that several mechanisms might be involved in the protective effects of LBP. Firstly, LBP restored cell proliferation in the hippocampus of SOC-treated rats. In the group that just received LBP but no SOC, there was also increased cell proliferation compared to the control groups. In addition, LBP also increased the number of newborn neurons in the dentate gyrus of the hippocampus, which indicated enhanced neurogenesis. Although there was no change in the number of immature neurons, the length of dendrites was markedly increased in the LBP + SOC group compared to the SOC group. Secondly, the LBP also attenuated SOC-induced oxidative stress, the levels of GP-x were marked increased, and that of MDA was reduced in LBP + SOC group compared to the SOC group. Thirdly, LBP has no effects on SOC-induced reduction of acetylcholinesterase (AChE) and the levels of brain-derived neurotrophic factor (BDNF) and insulin-like growth factor-1 (IGF-1) in the hippocampus. However, it did have antiapoptotic effects in the hippocampus, but the mechanism is uncertain (Chen et al. 2014).

Apart from animal studies, one human study also provided additional support to the beneficial effects of Wolfberry. β-carotene is one of the major components in Wolfberry. It is a precursor of vitamin A and it provides strong antioxidative effects. Every 100 g of Wolfberry contains 19.6 mg of carotene (Qi and Li 1981). In a randomized trial called "the Physicians' Health Study II (PHSII)," subjects who took β-carotene supplement (50 mg, alternate days) were followed up. For those who had short-term treatment of β-carotene (mean treatment duration = 1 year), they had similar cognition compared to the placebo group. However, for those who had received long-term treatment of β-carotene (mean treatment duration = 18 years), the mean global score for a group of cognition tests was significantly higher compared to the placebo group. The authors concluded that "the long-term β-carotene treatment was comparable to delaying cognitive aging by 1 to 1.5 years."

7.5 Concluding Remarks

Increasing lines of evidence support that AD is a multifactorial disease; therefore, its prevention or treatment should also be based on a multi-target approach. We have summarized how Wolfberry antagonizes various AD-risk factors in different studies. Moreover, Wolfberry has also been confirmed to improve cognition in animal models of AD. Human studies also suggest the potential role of Wolfberry in delaying aging-related cognitive impairment and improvement of general health. It is hoped that more in-depth and large-scale studies can be conducted in the near future to confirm the beneficial effects of Wolfberry on the aging population or dementia patients.

Acknowledgement The work is supported by the Macau Science and Technology Development Fund (053/2013/A2) to Y.S.H., HKU Alzheimer's Disease Research Network under Strategic Research Theme of Healthy Aging, Azalea (1972) Endowment Fund, and generous donation from Ms Kit-Wan Chow to RCCC.

References

Ahmed N, Ahmed U, Thornalley PJ, Hager K, Fleischer G, Munch G. Protein glycation, oxidation and nitration adduct residues and free adducts of cerebrospinal fluid in Alzheimer's disease and link to cognitive impairment. J Neurochem. 2005;92:255–63.

Alzheimer's disease association. 2012 Alzheimer's disease facts and figures. Alzheimers Dement. 2012;8:131–68.

Amagase H, Nance DM. Lycium barbarum increases caloric expenditure and decreases waist circumference in healthy overweight men and women: pilot study. J Am Coll Nutr. 2011;30:304–9.

Anton B, Vitetta L, Cortizo F, Sali A. Can we delay aging? The biology and science of aging. Ann N Y Acad Sci. 2005;1057:525–35.

Artola A, Kamal A, Ramakers GM, Biessels GJ, Gispen WH. Diabetes mellitus concomitantly facilitates the induction of long-term depression and inhibits that of long-term potentiation in hippocampus. Eur J Neurosci. 2005;22:169–78.

Barnes DE, Yaffe K. The projected effect of risk factor reduction on Alzheimer's disease prevalence. Lancet Neurol. 2011;10:819–28.

Brookmeyer R, Johnson E, Ziegler-Graham K, Arrighi HM. Forecasting the global burden of Alzheimer's disease. Alzheimers Dement. 2007;3:186–91.

Butters MA, Becker JT, Nebes RD, Zmuda MD, Mulsant BH, Pollock BG, Reynolds CF III. Changes in cognitive functioning following treatment of late-life depression. Am J Psychiatry. 2000;157:1949–54.

Cavallucci V, D'Amelio M, Cecconi F. Abeta toxicity in Alzheimer's disease. Mol Neurobiol. 2012;45:366–78.

Chen W, Cheng X, Chen J, Yi X, Nie D, Sun X, Qin J, Tian M, Jin G, Zhang X. Lycium barbarum polysaccharides prevent memory and neurogenesis impairments in scopolamine-treated rats. PLoS One. 2014;9:e88076.

Cheng D, Kong H. The effect of Lycium barbarum polysaccharide on alcohol-induced oxidative stress in rats. Molecules. 2011;16:2542–50.

Craft S, Peskind E, Schwartz MW, Schellenberg GD, Raskind M, Porte D Jr. Cerebrospinal fluid and plasma insulin levels in Alzheimer's disease: relationship to severity of dementia and apolipoprotein E genotype. Neurology. 1998;50:164–8.

Cui B, Liu S, Lin X, Wang J, Li S, Wang Q, Li S. Effects of Lycium barbarum aqueous and ethanol extracts on high-fat-diet induced oxidative stress in rat liver tissue. Molecules. 2011;16:9116–28.

Debette S, Markus HS. The clinical importance of white matter hyperintensities on brain magnetic resonance imaging: systematic review and meta-analysis. BMJ. 2010;341:c3666.

Douaud G, Refsum H, de Jager CA, Jacoby R, Nichols TE, Smith SM, Smith AD. Preventing Alzheimer's disease-related gray matter atrophy by B-vitamin treatment. Proc Natl Acad Sci U S A. 2013;110:9523–8.

Harman D. Alzheimer's disease pathogenesis: role of aging. Ann N Y Acad Sci. 2006;1067:454–60.

Hebert LE, Beckett LA, Scherr PA, Evans DA. Annual incidence of Alzheimer disease in the United States projected to the years 2000 through 2050. Alzheimer Dis Assoc Disord. 2001;15:169–73.

Ho YS, So KF, Chang RC. Drug discovery from Chinese medicine against neurodegeneration in Alzheimer's and vascular dementia. Chin Med. 2011;6:15.

Ho YS, Yang X, Lau JC, Hung CH, Wuwongse S, Zhang Q, Wang J, Baum L, So KF, Chang RC. Endoplasmic reticulum stress induces tau pathology and forms a vicious cycle: implication in Alzheimer's disease pathogenesis. J Alzheimers Dis. 2012;28:839–54.

Ho YS, Yu MS, Lai CS, So KF, Yuen WH, Chang RC. Characterizing the neuroprotective effects of alkaline extract of Lycium barbarum on beta-amyloid peptide neurotoxicity. Brain Res. 2007;1158:123–34.

Ho YS, Yu MS, Yang XF, So KF, Yuen WH, Chang RC. Neuroprotective effects of polysaccharides from wolfberry, the fruits of Lycium barbarum, against homocysteine-induced toxicity in rat cortical neurons. J Alzheimers Dis. 2010;19:813–27.

Ho YS, Yu MS, Yik SY, So KF, Yuen WH, Chang RC. Polysaccharides from wolfberry antagonizes glutamate excitotoxicity in rat cortical neurons. Cell Mol Neurobiol. 2009;29:1233–44.

Holliday R. Aging is no longer an unsolved problem in biology. Ann N Y Acad Sci. 2006;1067:1–9.

Hooshmand B, Polvikoski T, Kivipelto M, Tanskanen M, Myllykangas L, Erkinjuntti T, Makela M, Oinas M, Paetau A, Scheltens P, van Straaten EC, Sulkava R, Solomon A. Plasma homocysteine, Alzheimer and cerebrovascular pathology: a population-based autopsy study. Brain. 2013;136:2707–16.

Hynd MR, Scott HL, Dodd PR. Glutamate-mediated excitotoxicity and neurodegeneration in Alzheimer's disease. Neurochem Int. 2004;45:583–95.

Jimenez-Palomares M, Ramos-Rodriguez JJ, Lopez-Acosta JF, Pacheco-Herrero M, Lechuga-Sancho AM, Perdomo G, Garcia-Alloza M, Cozar-Castellano I. Increased Aβ production prompts the onset of glucose intolerance and insulin resistance. Am J Physiol Endocrinol Metab. 2012;302:E1373–80.

Jing L, Cui G, Feng Q, Xiao Y. Evaluation of hypoglycemic activity of the polysaccharides extracted from Lycium barbarum. Afr J Tradit Complement Altern Med. 2009;6:579–84.

Jorm AF. History of depression as a risk factor for dementia: an updated review. Aust N Z J Psychiatry. 2001;35:776–81.

Kamal A, Biessels GJ, Gispen WH, Ramakers GM. Synaptic transmission changes in the pyramidal cells of the hippocampus in streptozotocin-induced diabetes mellitus in rats. Brain Res. 2006;1073–1074:276–80.

Knegtering H, Eijck M, Huijsman A. Effects of antidepressants on cognitive functioning of elderly patients. A review. Drugs Aging. 1994;5:192–9.

Kopf D, Frolich L. Risk of incident Alzheimer's disease in diabetic patients: a systematic review of prospective trials. J Alzheimers Dis. 2009;16:677–85.

Korczyn AD. Why have we failed to cure Alzheimer's disease? J Alzheimers Dis. 2012;29:275–82.

Kulstad JJ, Green PS, Cook DG, Watson GS, Reger MA, Baker LD, Plymate SR, Asthana S, Rhoads K, Mehta PD, Craft S. Differential modulation of plasma beta-amyloid by insulin in patients with Alzheimer disease. Neurology. 2006;66:1506–10.

Leonard BE, Myint A. Inflammation and depression: is there a causal connection with dementia? Neurotox Res. 2006;10:149–60.

Li XM, Ma YL, Liu XJ. Effect of the Lycium barbarum polysaccharides on age-related oxidative stress in aged mice. J Ethnopharmacol. 2007;111:504–511.

Lin NC, Lin JC, Chen SH, Ho CT, Yeh AI. Effect of Goji (Lycium barbarum) on expression of genes related to cell survival. J Agric Food Chem. 2011;59:10088–96.

Luo Q, Cai Y, Yan J, Sun M, Corke H. Hypoglycemic and hypolipidemic effects and antioxidant activity of fruit extracts from Lycium barbarum. Life Sci. 2004;76:137–49.

McIlroy SP, Dynan KB, Lawson JT, Patterson CC, Passmore AP. Moderately elevated plasma homocysteine, methylenetetrahydrofolate reductase genotype, and risk for stroke, vascular dementia, and Alzheimer disease in Northern Ireland. Stroke. 2002;33:2351–6.

Morris MS. Homocysteine and Alzheimer's disease. Lancet Neurol. 2003;2:425–8.

Murphy TH, Schnaar RL, Coyle JT. Immature cortical neurons are uniquely sensitive to glutamate toxicity by inhibition of cystine uptake. FASEB J. 1990;4:1624–33.

Ownby RL, Crocco E, Acevedo A, John V, Loewenstein D. Depression and risk for Alzheimer disease: systematic review, meta-analysis, and metaregression analysis. Arch Gen Psychiatry. 2006;63:530–8.

Paul Hsu CH, Nance DM, Amagase H. A meta-analysis of clinical improvements of general well-being by a standardized Lycium barbarum. J Med Food. 2012;15:1006–14.

Poon DC, Ho YS, Chiu K, Chang RC. Cytokines: how important are they in mediating sickness? Neurosci Biobehav Rev. 2013;37:1–10.

Qi ZS, Li SF. Determination of the chemical composition of Fructus Lycii. 1981. pp. 67–72.

Querfurth HW, LaFerla FM. Alzheimer's disease. N Engl J Med. 2010;362:329–44.

Shen C, Chen Y, Liu H, Zhang K, Zhang T, Lin A, Jing N. Hydrogen peroxide promotes Aβ production through JNK-dependent activation of gamma-secretase. J Biol Chem. 2008;283;17721–30.

Starkstein SE, Mizrahi R, Power BD. Depression in Alzheimer's disease: phenomenology, clinical correlates and treatment. Int Rev Psychiatry. 2008;20:382–88.

Tang T, He B. Treatment of d-galactose induced mouse aging with lycium barbarum polysaccharides and its mechanism study. Afr J Tradit Complement Altern Med. 2013;10:12–7.

Wong GT, Chang RC, Law AC. A breach in the scaffold: the possible role of cytoskeleton dysfunction in the pathogenesis of major depression. Ageing Res Rev. 2013;12:67–75.

Wrighten SA, Piroli GG, Grillo CA, Reagan LP. A look inside the diabetic brain: contributors to diabetes-induced brain aging. Biochim Biophys Acta. 2009;1792:444–53.

Wuwongse S, Chang RC, Law AC. The putative neurodegenerative links between depression and Alzheimer's disease. Prog Neurobiol. 2010;91:362–75.

Wuwongse S, Cheng SS, Wong GT, Hung CH, Zhang NQ, Ho YS, Law AC, Chang RC. Effects of corticosterone and amyloid-beta on proteins essential for synaptic function: implications for depression and Alzheimer's disease. Biochim Biophys Acta. 2013;1832:2245–56.

Yu MS, Ho YS, So KF, Yuen WH, Chang RC. Cytoprotective effects of Lycium barbarum against reducing stress on endoplasmic reticulum. Int J Mol Med. 2006;17:1157–61.

Yu MS, Lai CS, Ho YS, Zee SY, So KF, Yuen WH, Chang RCC. Characterization of the effects of anti-aging medicine Fructus lycii on beta-amyloid peptide neurotoxicity. Int J Mol Med. 2007;20:261–8.

Yu MS, Leung SK, Lai SW, Che CM, Zee SY, So KF, Yuen WH, Chang RC. Neuroprotective effects of anti-aging oriental medicine Lycium barbarum against beta-amyloid peptide neurotoxicity. Exp Gerontol. 2005;40:716–27.

Zhang E, Yau SY, Lau BW, Ma H, Lee TM, Chang RC, So KF. Synaptic plasticity, but not hippocampal neurogenesis, mediated the counteractive effect of wolfberry on depression in rats (1). Cell Transplant. 2012;21:2635–49.

Zhang Q DUX, Xu Y, Dang L, Xiang L, Zhang J. The effects of Gouqi extracts on Morris maze learning in the APP/PS1 double transgenic mouse model of Alzheimer's disease. Exp Ther Med. 2013;5:1528–30.

Zhu J, Liu W, Yu J, Zou S, Wang J, Yao W, Gao X. Characterization and hypoglycemic effect of a polysaccharide extracted from the fruit of Lycium barbarum L. Carbohydr Polym. 2013;98:8–16.

Chapter 8
Prosexual Effects of *Lycium Barbarum*

**Benson Wui-Man Lau, Mason Chin-Pang Leung, Kai-Ting Po,
Raymond Chuen-Chung Chang and Kwok-Fai So**

Abstract *Lycium barbarum,* also known as wolfberry, has been commonly used as an oriental herb in China for a long period. Wolfberry is suggested to be antiaging and used to maintain health of kidneys, liver, and the eyes. In the practice of traditional Chinese medicine, wolfberry was also prescribed for patients who suffered from dysfunctions of sexual desire. Although it has been put into application for a long history, the underlying mechanism is still elusive. In the past decades, increasing lines of evidence support the roles of wolfberry in promoting male sexual functioning. The benefit of wolfberry was shown on different aspects including preventing the reproductive tissues from oxidative insult, improving motility of sperm, maintaining a testosterone level, and promoting sexual performance of the hemicastrated rats. This chapter discusses on the recent research findings, both from bench and bedside, which support the prosexual function of wolfberry, and discusses the potential clinical application of wolfberry on the treatment of sexual behavior.

B. W.-M. Lau (✉) · R. C.-C. Chang
LKS Faculty of Medicine, Department of Anatomy, The University of Hong Kong,
Pokfulam, Hong Kong
e-mail: benson.lau@polyu.edu.hk

B. W.-M. Lau · M. C.-P. Leung · K.-T. Po
Department of Rehabilitation Sciences, The Hong Kong Polytechnic University,
Kowloon, Hong Kong, People's Republic of China

K.-F. So
LKS Faculty of Medicine, Department of Ophthalmology, The University of Hong Kong,
Pokfulam, Hong Kong, People's Republic of China

R. C.-C. Chang · K.-F. So
The State Key Laboratory of Brain and Cognitive Science, The University of Hong Kong,
Pokfulam, Hong Kong, People's Republic of China

R. C.-C. Chang · K.-F. So
LKS Faculty of Medicine, Research Centre of Heart, Brain, Hormone and Healthy Aging,
The University of Hong Kong, Pokfulam, Hong Kong, People's Republic of China

R. C.-C. Chang · K.-F. So
GMH institute of Central nervous System Regeneration, Guangdong Key Laboratory of Brain
Function and Diseases, Jinan University, Guangzhou, People's Republic of China

© Springer Science+Business Media Dordrecht 2015
R. C-C. Chang, K-F. So (eds.), *Lycium Barbarum and Human Health*,
DOI 10.1007/978-94-017-9658-3_8

Keywords Wolfberry · *Lycium barbarum* · Sexual behaviors · Mating · Reproduction · Aphrodisiac

The prosexual and fertility effect of wolfberry was first described by the Chinese herbalist Li Shizhen (Wang et al. 2002). While in traditional Chinese medicine (TCM) practice, wolfberry is commonly prescribed for sexual dysfunction, the beneficial effect of wolfberry on reproduction provided by scientific methodology has just investigated in the past two decades.

8.1   Beneficial Effect of Wolfberry on Sexual Behavior

The first evidence which supports the prosexual effect of wolfberry polysaccharide on live animals was reported by Luo et al. (2006). *Lycium barbarum* polysaccharide (LBP) treatment for 22 days was shown to attenuate the impaired copulatory performance caused by hemicastration in rats, which is shown by markedly shortened penis erection latency, mounts latency, and increased successful mount percentage. Interestingly, the sexual performance of rats with LBP treatment is even slightly better than that of normal rats. In parallel to the sexual behavior, sex hormones in plasma and the weight of accessory reproductive organ are also maintained in hemicastrated rats treated with LBP. To elucidate the protective effect at cellular level, the authors also tested the protective effect of LBP on testicular tissues. When being subjected to hyperthermia, rat testis tissues were found to have decreased weight and superoxide dismutase (SOD) activity, which is associated with a structural damage of the seminiferous tubules. Pretreatment with LBP before the heat exposure was able to prevent the abovementioned damage on the testicular tissue. Furthermore, treatment with LBP in vitro also prevented the deoxyribonucleic acid (DNA) damage caused by H_2O_2. These findings support the traditional Chinese viewpoint that wolfberry is an aphrodisiac agent and may facilitate fertility by scientific evidence.

Apart from TCM practice, wolfberry was also adapted as a remedy to improve sexual functioning in Korea (Sohn et al. 2008). An herbal formulation consists of the seeds of wolfberry and other herbs were tested for its effect on penile erection. After treatment for 4 weeks, the intracavernosal pressure was found to increase markedly in medicated animals, which indicate that erection was promoted. Simultaneously, an expression of nNOS and eNOS (neuronal and endothelial nitric oxide synthase, respectively), which generate nitric oxide as a vasodilator, was also found to be upregulated by the medication. These observations suggested that the herbal mixture may promote male copulatory function by promoting penile erection.

A study conducted by Lau et al. tested whether the LBP could counteract the sexual-suppressing effect of high dose corticosterone (Lau et al. 2012). As several studies suggested that neurogenesis (i.e., production of new functional neurons)

may be an essential mediator of sexual behavior and newborn neurons may have an important role in the regulation of sexual behavior (Lau et al. 2011; Leuner et al. 2010), the authors examined whether neurogenesis takes roles in the prosexual effect of LBP and the association between LBP treatment, sexual behavior, and neurogenesis. The results of the study indicated that LBP facilitated male sexual behavior by significantly increased copulatory efficiency (CE) and ejaculation frequency (EF), and decreased ejaculation latency (EL). LBP at dosage of 1 mg/kg showed the most significant effect than a higher dosage of 10 mg/kg.

Furthermore, LBP also significantly reversed the inhibited sexual behavior induced by corticosterone in intromission latency (IL) and CE and reversed suppressed neurogenesis induced by corticosterone in terms of number of BrdU-positive cells in subventricular zone (SVZ) and hippocampus. LBP also promotes the neuronal differentiation of neural precursor cells as shown by DCX staining. Moreover, the study further revealed that sexual performance is correlated with neurogenesis in SVZ and hippocampus and the direct relationship is examined by comparing BrdU-positive cells and sexual behavior between blocking neurogenesis in LBP treatment group and control group. After blocking neurogenesis, the rats showed impaired sexual performance, which indicate the necessity of neurogenesis in sexual functioning. This study demonstrated the enhancing effect of LBP on male sexual behaviors in rats, the association between neurogenesis and the causal relationship between neurogenesis and sexual behavior. As newborn neurons may have an important role in the regulation of sexual behavior, this study suggests that LBP may promote sexual behavior through the regulation of neurogenesis.

A randomized controlled clinical study was conducted to investigate the effect of a drink which was prepared from wolfberry ("Goji Juice GoChi") on general health of human subjects (Amagase and Nance 2008). The drink was prepared from fresh wolfberry, with a daily serving which is equivalent to at least 150 g of fresh wolfberry. Questionnaires were used to evaluate subjective well-being after consuming the drink for 14 days. In comparing to the placebo group, the subjects who had the wolfberry drink responded with a significantly higher energy level, feeling of health and exercise performance, and reduction in fatigue and stress. Interestingly, more than half of the female subjects consumed the juice reported decrease in pain during their mentrual period, and a few subjects in the treatment groups stated they had increased sexual ability and activity. Since these aspects on reproduction were not the main foci of the randomized controlled trial (RCT), no dedicated instrument and analysis were used to evaluate the difference between the treatment and placebo groups. Another study reported by the same research group indicated that after chronic consumption of the wolfberry drink for 30 days, the in vivo antioxidant markers including SOD and glutathione peroxidase (GSH-Px) in the plasma were upregulated (Amagase et al. 2009), which exposed the antioxidant properties of the wolfberry drink on human. As oxidative stress may be one of the underlying mechanisms of sexual inhibition and infertility, the antioxidant action of wolfberry drink in human may be linked to the prosexual effect reported by the previous study. Nevertheless, the effect on the subjective feelings on human sexual function would be worthwhile for future RCT experiment. As increasing lines of evidence showed the prosexual effect of wolfberry in laboratory animals, further studies on the potential reproductive effect of wolfberry on human would be valuable for the translation of laboratory studies to clinical applications.

8.2   Protective Effect of Wolfberry on Reproductive Tissues

Since wolfberry has been shown to exert its effect by its antioxidant activity, it was hypothesized that antioxidation is a mechanism that underlies the beneficial effect of wolfberry on reproductive function. As a by-product of intracellular respiration, reactive oxygen species (ROS) are reactive molecules which may damage the reproductive tissues if in excess amount (Balaban et al. 2005). ROS species including hydroxyl ions, superoxide, and nitric oxides are generated in the testis during steroidogenesis and spermatogenesis (Mathur and D'Cruz 2011). While an appropriate level of ROS is required for the normal functioning of the testis tissue and sperms, an excessive amount of ROS is detrimental to the sperms as the sperms are highly sensitive to peroxidation due to the rich content of polyunsaturated fatty acids (PUFA) on the plasma membrane. A proper balance between the ROS and antioxidants is thus needed to maintain the functioning of the reproductive tissues and sperms. As antioxidants were found in various edible plants (Clément et al. 2012), it is suggested that plant-derived feed supplements may inhibit the oxidative stress in livestocks and may promote both fertility and reproductive functions. One of the food supplements that received increasing attention is wolfberry.

Different studies have investigated the protective effect of wolfberry on reproductive tissues by both in vivo and in vitro methods, and tried to determine whether wolfberry exerts its effect through its antioxidant properties. The protective effect of wolfberry polysaccharides on testicular tissue was firstly shown in a tissue culture study (Wang et al. 2002). When seminiferous epithelium was cultured for a prolonged period and at a relatively high temperature, structural degradation was induced in terms of increase in intercellular space and appearance of multinucleated spermatids, which is due to the impaired spermatogenesis. Inclusion of wolfberry polysaccharide in the culture medium reduced the structural damage caused by the lengthened culture time and hyperthermia. The spermatozoa were also found to be more motile when treated with LBP. The reduced structural damage is accompanied with a decrease in apoptotic rate of epithelium cells and reduction in oxidative stress (illustrated by ultraviolet C light induced lipid peroxidation and superoxide induced cytochrome c reduction). The authors suggest that the protective effect of LBP may be due to its antioxidant properties.

In another study conducted by Zhang et al. showed that LBP could protect the spermatogenesis in testis caused by Bisphenol A (BPA) (Zhang et al. 2013). BPA is a commonly found monomer in plastic wares and widely used in adhesive and dental fillings (Podlipna and Cichna-Markl 2006). Despite its wide range of applications, BPA was found to decrease the weight of testis, induce apoptosis in spermatogenic tissues, decrease testosterone levels and increase the rate of infertility in rats (Li et al. 2009; Podlipna and Cichna-Markl 2006; Xiao et al. 2011). Interestingly, LBP was shown to prevent the retardation of reproductive function in rats caused by BPA. First, LBP was shown to increase the weight of testis and epididymis in BPA-treated rats, although not to a level comparable to the normal controls. Second, LBP was shown to prevent the decrease in the level of sex hormones caused by BPA. Third, the proapoptotic marker, Bax was found to decrease after LBP treatment. Finally, oxidative stress was also suppressed by the LBP. These data suggest that LBP would mitigate the damage of testicular tissues from BPA by suppressing apoptosis and oxidative stress, while further exploration of the fertility of the rats after BPA and LBP treatment would support the use of wolfberry in reproductive function against toxic agents on reproductive system.

Another study explored whether pretreatment with LBP could ameliorate the detrimental effect on reproductive system caused by doxorubicin (DOX), an antitumor drug used in the treatment of solid and hematological tumors (Xin et al. 2012). Aside from its side effect on cardiopulmonary and excretory system (Minotti et al. 2004), DOX also causes toxicity on testicular tissue and spermatozoa manifested in decreasing quantity and motility of sperm, increasing the rate of abnormal sperm production and increasing apoptosis in the spermatogenesis process. Being similar to the abovementioned studies, the authors hypothesized that oxidative stress is a major underlying mechanism of the disturbed reproductive function. Evidences indicated that the weight of the testis and epididymis in DOX-treated male rats was preserved by LBP.

Morphologically, degenerative changes including depletion of germ cells, irregular seminiferous tubules and scarcity of spermatogonia were caused by DOX, which could be prevented by LBP. While LBP does not show any effect on sperm quantity and quality in healthy rats, it effectively prevented the decline of sperm quantity and quality which is induced by DOX. Furthermore, the increased oxidative stress in testicular tissues caused by DOX, indicated by the level of malondialdehyde (by-product of lipid peroxidation) and GSH-peroxidase (a scavenger of free radicals), was attenuated by LBP. Again, the evidence suggests that LBP may be a potential adjunct therapy for the protection of reproductive organs through the regulation of oxidative stress.

Different lines of evidence showed that ionizing irradiation has damaging effects in male reproductive system (Bonde 2010) and testicular spermatogenic cells were highly sensitive to ionizing irradiation. Cancer patients receiving radiotherapy or people working with radioactive substances often had infertility and sexual dysfunction (Hasegawa et al. 1997). A study attempted to elucidate whether LBP could protect the reproductive tissues from ionizing irradiation-induced reproductive cell damage (Luo et al. 2011). Male rats were exposed to various levels of ionizing irradiation and then cotreated with LBP. After the experiment, sperm count, their motility, erection latency, sexual behavior, serum hormone, DNA damage in testicular cells, and protein content of testicular tissues were tested. The result showed that in irradiated animals, sperm count and their motility were significantly reduced. Moreover, erection latency, mounting, and ejaculation latencies were extended, serum testosterone levels was lower and DNA damage in testicular cells were observed after exposing to ionizing irradiation. The longer the exposure to ionizing irradiation, the greater the deleterious effect will be found. Cotreatment with LBP was shown to significantly alleviate the detrimental effect on the reproductive tissues. The author suggested that LBP could repair the damage in testicular cells caused by ionizing irradiation, regulate the serum testosterone level, and protect testicular cells against deleterious effects of free radicals caused by ionizing irradiation. A subsequent study conducted by the same group showed that LBP upregulates the expression of antiapoptotic protein Bcl-2, downregulated proapoptotic Bax and maintain the mitochondrial membrane potential of testicular tissues (Luo et al. 2014), which suggests that LBP may have antiapoptotic effects against irradiation.

Owing to the traditional viewpoint that wolfberry is aphrodisiac and its availability is at low cost, the potential profertility effect of wolfberry in female has drawn the attention of research groups as early as the 1970s (Suzuki et al. 1972). In a study conducted by Suzuki and coworkers, intravenous injection of crude water-soluble extract of *Lycium Chinense*, another species of wolfberry or goji berry, induced ovulation in rabbits (Suzuki et al. 1972). However, due to the lack of an appropriate control group in the abovementioned study, whether the ovulation is induced by the wolfberry extract or due to other confounds remains to be determined.

Huang et al. studied the effect of LBP on in vitro maturation of the female gamete (Huang et al. 2008). Interestingly, when frozen oocyte with cryoprotection was thawed and treated with LBP in culture conditions, the maturation rate was significantly higher than the traditional sucrose medium. However, while the maturation-promoting action was attributed to the influence on solution viscosity and osmolality, it is unknown of whether the effect is specific to the wolfberry.

In a case study, concentrated herbal extracts (Zuo-gui-wan), which contains cooked wolfberry and other herbs such as dogwood fruit and cyathula root, was used to treat a woman with premature ovarian failure and secondary amenorrhea (Chao et al. 2003). After 3 months of therapy, ovulation return and the woman conceived successfully.

Interestingly, wolfberry was found to benefit not only the adult animals but also offsprings when they are exposed to it during the gestation period (Feng et al. 2010). It was found that prenatal stress resulted in a significant decrease in cognitive function in female offspring, which could be prevented significantly by pretreatment of mother rats with milk-based wolfberry. After the pretreatment with wolfberry polysaccharide for 2 weeks, female rats were allowed to mate. Then, the pregnant female subjects were restrained by a transparent plastic tube on days 14–20 of pregnancy three times daily from 45 min to 1 h. The female offspring were subsequently tested at 1 month of age. Morris Water Maze was used to test the spatial memory and the offspring rats were sacrificed to test the oxidative brain mitochondrial damage. The result showed that prenatal restraint stress induced memory and learning deficiency of female offspring, but not in male offspring, while middle and high doses wolfberry pretreatment significantly reduced the impairment. In vitro studies by the authors showed that wolfberry dose dependently scavenged hydroxyl and superoxide radicals and inhibited ascorbic acid-induced dysfunction in brain tissue and tissue mitochondria.

When comparing to male, the effect of wolfberry on female reproductive function and offspring remains unclear and only a few limited reports studied the effect. There is a lack of supportive scientific evidence on the effect of wolfberry on female sexual behavior and reproduction. However, as it was shown that wolfberry could increase the levels of plasma sex hormones and protect the male gametes, it is likely that female reproductive system may be benefited by the herb. The supporting evidence from empirical studies will definitely strengthen the clinical usage of wolfberry in practice, and even as a food supplement.

In conclusion, evidence of *L. barbarum*'s prosexual effect has been found at different levels of organisms, namely: molecular level, biochemistry level, cellular level, tissue level, behavior level. Also, it improves the general well-being of humans. We have summarized all these effects in Table 8.1.

Conclusion

Apart from protecting the DNA and alleviating the reproductive tissue damage due to hypothermia, oxidative stress, and radiation, the fruit also protects testis from damage by environment pollutant like BPA. Female fertility is also restored by a herbal medicine containing wolfberry. The fruit also shows protective effect to the cognitive function of offspring with prenatal stress. Different pieces of evidence have shown that *L. barbarum* would be an all-round aphrodisiac agent that not only improves the sexual function of male individual and the fertility of female, but also the general health of both male and female. These beneficial effects imply *L. barbarum* would be a potent functional food. Nevertheless, the drug interaction between the fruit and common drugs, especially drugs for chronic diseases, still need to be clarified, in order to utilize wolfberry as an adjunctive therapy for sexual dysfunctions. Yet, further studies may be needed to determine the therapeutic dosage of wolfberry as the quality control of the fruit and the active components are still being elusive.

Table 8.1 Summary of the prosexual effect of Wolfberry

Sex of subjects	Reproductive stages	Species or experimental model	Effect of wolfberry component	Reference
Male ♂	Mating	Rats, hemicastrated	Shortened penis erection latency, mounts latency	Luo et al. 2006
			Increased successful mount percentage	
		Rats	Promote penile erection	Sohn et al. 2008
			Penile expression level of eNOS/nNOS increased	
		Rats, treated with high dose corticosterone	Increase CE and ejaculation frequency	Lau et al. 2012
			Decrease ejaculation latency	
			Reverse suppressed neurogenesis in subventricular zone and hippocampus	
		Rats, hemicastrated	Improve CE of hemicastracted rats	Luo et al. 2006
		Mouse testicular tissue	Prevent the structural damage of the seminiferous tubules	
			Prevent DNA damage caused by H_2O_2	
		Mouse seminiferous epithelium	Reduced structural damage of seminiferous epithelium	Wang et al. 2002
			More motile spermatozoa	
		Rats, with BPA treatment	*Prevent detrimental effect caused by BPA in terms of*: increase the weight of testis and epididymis	Zhang et al. 2013
			Prevent the decrease in the level of sex hormones	
			Decreased level of proapototic marker, e.g., Bax	
			Suppressed oxidative stress	
		Rats, treated with DOX	*Prevents detrimental effect caused by DOX in terms of*: preserved weight of the testis and epididymis—prevent degenerative changes (depletion of germ cells, irregular seminiferous tubules and scarcity of spermatogonia)	Xin et al. 2012
			Prevent the decline of sperm quantity and quality	
			Attenuate the increased oxidative stress in testicular tissues	

Table 8.1 (continued)

Sex of subjects	Reproductive stages	Species or experimental model	Effect of wolfberry component	Reference
		Rats, exposed to ionizing radiation	*Prevents detrimental effect caused by ionizing radiation in terms of:* repair the damage in testicular cells	Luo et al. 2011
			Regulate the serum testosterone level and	
			Maintain histological integrity of testicular tissue	
			May have antiapoptotic effects against irradiation	Luo et al. 2014
Female♀	Mating	Rabbits	Induce ovulation after injection of crude extract of wolfberry	Suzuki et al. 1972
	Mating	Frozen porcine oocyte	Increase maturation rate	Huang et al. 2008
	Gestation	Human, single case	*A woman with premature ovarian failure and secondary amenorrhea had ovulation returned and conceived after 3 months of therapy*	Chao et al. 2003
	Offspring	Rats, offspring	Reduce the impairment in memory and learning deficiency of female offspring which were subjected to prenatal stress	Feng et al. 2010
Both sexes	Mating	Normal human	*Drinking wolfberry juice showed*	Amagase and Nance 2008, 2009
			Subjective report of increased sexual ability and activity	
			Elevated level of in vivo antioxidant markers including SOD and GSH-Px in plasma	

eNOS endothelial nitric oxide synthase, *nNOS* neuronal nitric oxide synthase, *CE* copulatory efficiency, *DNA* deoxyribonucleic acid, *BPA* bisphenol A, *DOX* doxorubicin, *SOD* superoxide dismutase, *GSH-Px* glutathione peroxidase

Reference

Amagase H, Nance DM. A randomized, double-blind, placebo-controlled, clinical study of the general effects of a standardized Lycium barbarum (Goji) juice, GoChi. J Altern Complement Med. 2008;14:403–12. doi:10.1089/acm.2008.0004.

Amagase H, Sun B, Borek C. Lycium barbarum (goji) juice improves in vivo antioxidant biomarkers in serum of healthy adults. Nutr Res. 2009;29:19–25. doi:10.1016/j.nutres.2008.11.005.

Balaban RS, Nemoto S, Finkel T. Mitochondria, oxidants, and aging. Cell. 2005;120:483–95. doi:10.1016/j.cell.2005.02.001.

Bonde JP. Male reproductive organs are at risk from environmental hazards. Asian J Androl. 2010;12:152–6. doi:10.1038/aja.2009.83.

Chao S-L, Huang L-W, Yen H-R. Pregnancy in premature ovarian failure after therapy using Chinese herbal medicine. Chang Gung Med J. 2003;26:449–52.

Clément C, Witschi U, Kreuzer M. The potential influence of plant-based feed supplements on sperm quantity and quality in livestock: a review. Anim Reprod Sci. 2012;132:1–10. doi:10.1016/j.anireprosci.2012.04.002.

Feng Z, Jia H, Li X, Bai Z, Liu Z, Sun L, Zhu Z, Bucheli P, Ballèvre O, Wang J, Liu J. A milk-based wolfberry preparation prevents prenatal stress-induced cognitive impairment of offspring rats, and inhibits oxidative damage and mitochondrial dysfunction in vitro. Neurochem Res. 2010;35:702–11. doi:10.1007/s11064-010-0123-5.

Hasegawa M, Wilson G, Russell LD, Meistrich ML. Radiation-induced cell death in the mouse testis: relationship to apoptosis. Radiat Res. 1997;147:457–67.

Huang J, Li Q, Zhao R, Li W, Han Z, Chen X, Xiao B, Wu S, Jiang Z, Hu J, Liu L. Effect of sugars on maturation rate of vitrified-thawed immature porcine oocytes. Anim Reprod Sci. 2008;106:25–35. doi:10.1016/j.anireprosci.2007.03.023.

Lau BW-M, Yau S-Y, So K-F. Reproduction: a new venue for studying function of adult neurogenesis? Cell Transplant. 2011;20:21–35. doi:10.3727/096368910X532765.

Lau BW-M, Lee JC-D, Li Y, Fung SM-Y, Sang Y-H, Shen J, Chang RC-C, So K-F. Polysaccharides from wolfberry prevents corticosterone-induced inhibition of sexual behavior and increases neurogenesis. PLoS One. 2012;7:e33374. doi:10.1371/journal.pone.0033374.

Leuner B, Glasper ER, Gould E. Sexual experience promotes adult neurogenesis in the hippocampus despite an initial elevation in stress hormones. PLoS One. 2010;5:e11597. doi:10.1371/journal.pone.0011597.

Li Y-J, Song T-B, Cai Y-Y, Zhou J-S., Song X, Zhao X, Wu X-L. Bisphenol A exposure induces apoptosis and upregulation of Fas/FasL and caspase-3 expression in the testes of mice. Toxicol Sci. 2009;108:427–36. doi:10.1093/toxsci/kfp024.

Luo Q, Li Z, Huang X, Yan J, Zhang S, Cai Y-Z. Lycium barbarum polysaccharides: protective effects against heat-induced damage of rat testes and H2O2-induced DNA damage in mouse testicular cells and beneficial effect on sexual behavior and reproductive function of hemicastrated rats. Life Sci. 2006;79:613–21. doi:10.1016/j.lfs.2006.02.012.

Luo Q, Cui X, Yan J, Yang M, Liu J, Jiang Y, Li J, Zhou Y. Antagonistic effects of Lycium barbarum polysaccharides on the impaired reproductive system of male rats induced by local subchronic exposure to 60Co-γ irradiation. Phytother Res. 2011;25:694–701. doi:10.1002/ptr.3314.

Luo Q, Li J, Cui X, Yan J, Zhao Q, Xiang C. The effect of Lycium barbarum polysaccharides on the male rats' reproductive system and spermatogenic cell apoptosis exposed to low-dose ionizing irradiation. J Ethnopharmacol. 2014;154(1):249–58. doi:10.1016/j.jep.2014.04.013.

Mathur PP, D'Cruz SC. The effect of environmental contaminants on testicular function. Asian J Androl. 2011;13:585–91. doi:10.1038/aja.2011.40.

Minotti G, Menna P, Salvatorelli E, Cairo G, Gianni L. Anthracyclines: molecular advances and pharmacologic developments in antitumor activity and cardiotoxicity. Pharmacol Rev. 2004;56:185–229. doi:10.1124/pr.56.2.6.

Podlipna D, Cichna-Markl M. Determination of bisphenol A in canned fish by sol—gel im-munoaffinity chromatography, HPLC and fluorescence detection. Eur Food Res Technol. 2006;224:629–34. doi:10.1007/s00217-006-0350-9.

Sohn DW, Kim HY, Kim SD, Lee EJ, Kim HS, Kim JK, Hwang SY, Cho Y-H, Kim SW. Elevation of intracavernous pressure and NO-cGMP activity by a new herbal formula in penile tissues of spontaneous hypertensive male rats. J. Ethnopharmacol. 2008;120:176–80. doi:10.1016/j.jep.2008.08.005.

Suzuki M, Osawa S, Hirano M. A Lycium chinense miller component inducing ovulation in adult female rabbits. Tohoku J Exp Med. 1972;106:219–31.

Wang Y, Zhao H, Sheng X, Gambino PE, Costello B, Bojanowski K. Protective effect of Fructus Lycii polysaccharides against time and hyperthermia-induced damage in cultured seminiferous epithelium. J Ethnopharmacol. 2002;82:169–75.

Xiao S, Diao H, Smith MA, Song X, Ye X. Preimplantation exposure to bisphenol A (BPA) af-fects embryo transport, preimplantation embryo development, and uterine receptivity in mice. Reprod Toxicol. 2011;32:434–41. doi:10.1016/j.reprotox.2011.08.010.

Xin Y-F, You Z-Q, Gao H, Zhou G-L, Chen Y-X, Yu J, Xuan Y-X. Protective effect of Lycium bar-barum polysaccharides against doxorubicin-induced testicular toxicity in rats. Phytother Res. 2012;26:716–21. doi:10.1002/ptr.3633.

Zhang C, Wang A, Sun X, Li X, Zhao X, Li S, Ma A. Protective effects of Lycium barbarum polysaccharides on testis spermatogenic injury Induced by bisphenol A in mice. Evid Based Complement Altern Med. 2013;2013:690808. doi:10.1155/2013/690808.

Chapter 9
Lycium Barbarum: Neuroprotective Effects in Ischemic Stroke

Amy CY Lo and Di Yang

Abstract Ischemic stroke is a leading cause of death worldwide, bringing about serious long-lasting disability and considerable social burden. Stroke patients suffer from various disabilities, including hemiplegia, dysesthesia, ataxia, and sometimes visual impairment. During an ischemic stroke, ischemia takes place due to blood flow interruption as a result of a cerebral artery blockade. The disruption in oxygen and glucose supply and subsequent reperfusion trigger a complex cascade of molecular events that eventually results in irreversible cell death in the affected brain area, affecting the functioning of the body. Treatment of ischemic stroke is important to alleviate the subsequent outcome. Yet, no ideal neuroprotective agents are available. Some research has turned to traditional medicine that has shown efficacy in animal models, making it an attractive option in the treatment of ischemic stroke. Another appealing alternative would be the prevention of ischemic stroke using traditional medicine, which may be beneficial for patients at high risk for ischemic stroke. One traditional medicine that shows promising results is *Lycium barbarum* (wolfberry, Goji, *Fructus Lycii*), an important traditional medicine in promoting health and longevity as well as a food supplement in the Western countries.

Keywords Ischemia-reperfusion injury · Middle cerebral artery occlusion · Neurological deficit · Cerebral infarct · Cerebral edema · Apoptosis · Blood-brain barrier · Aquaporin · GFAP · Pretreatment · Prophylaxis · Retina · Retinal ganglion cell

9.1 Introduction

Stroke is the second leading cause of death in the world, with more than 15 million sufferers while one third of them die every year. The survivors are left with various cruel sequelae, including hemiplegia, dysesthesia, ataxia, and even visual

A. C. Lo (✉) · D. Yang
Department of Ophthalmology, Li Ka Shing Faculty of Medicine,
The University of Hong Kong, 21 Sassoon Road,
Hong Kong, People's Republic of China
e-mail: amylo@hku.hk

© Springer Science+Business Media Dordrecht 2015
R. C-C. Chang, K-F. So (eds.), *Lycium Barbarum and Human Health,*
DOI 10.1007/978-94-017-9658-3_9

impairment such as hemianopia (Barber et al. 2001), seriously affecting their quality of life. As the world population ages, management of these subsequent disabilities pose an increasing burden to the whole society, making stroke a devastating cerebrovascular disease.

9.2 Stroke and Cerebral Ischemia: Pathophysiology and Mechanisms

There are two major categories of stroke according to its etiology, namely ischemic stroke and hemorrhagic stroke. Out of the two types of stroke, ischemic stroke accounts for approximately 80 % of all cases (Goldstein et al. 2006). During an ischemic stroke, there is cerebral artery blockade and therefore cerebral ischemia.

Cerebral ischemia occurs as a result of an obstruction in one of the major cerebral blood vessels, leading to reduced blood supply to its irrigation territory and therefore insufficient glucose and oxygen delivery to the affected area in the brain. An ischemic core and surrounding it the penumbra soon develop (Astrup et al. 1981; Olsen et al. 1983). Normally, the cells in the brain and nervous system require abundant blood supply in order to maintain proper neuronal function. Once the blood flow drops below 25 % of normal values during cerebral ischemia, neuronal activity declines rapidly. With the insufficient levels of oxygen and glucose, energy depletion results in a series of molecular events triggered by ischemia, namely ischemia cascade (Wityk and Stern 1994; Love 1999; Barber et al. 2001). These include disruption of ion homeostasis, depolarization, calcium channel dysfunction, excessive glutamate release, excitotoxicity, and release of free radical that rapidly result in cell death. These events happen rapidly in the infarct core where blood flow is most severely restricted (Barone et al. 2002; Doyle et al. 2008). Not only the neurons are affected; other cells such as glia and endothelial cells are also damaged.

When the blood supply to targeted tissues restores, reperfusion starts. It is widely accepted that reperfusion brings further damage to the ischemic tissues and this damage may be more serious than the initial ischemic injury. After the blood flow is resumed, more free radicals and reactive oxygen species are generated. The early events mentioned above are followed by late events such as inflammatory changes, apoptosis, and disruption of the blood-brain barrier that evolve over hours or even days (Dirnagl et al. 1999; Lin 2002; Lo et al. 2003), causing exaggerated cellular damage that eventually result in additional impairment to the tissue.

9.3 Stroke: Treatment Strategies

The key therapeutic strategy for patients with acute ischemic stroke mainly focuses on early revascularization in order to restore blood flow to the ischemic brain tissue at early stage. This helps protect the neurons from further damage and reduce the long-term disabilities associated with stroke attack. Present management includes pharmacologic thrombolysis, mechanical thrombectomy, and angioplasty. To date, the only approved medicine by the US Food and Drug Administration (FDA) for patients with acute ischemic stroke is recombinant tissue plasminogen activator (rtPA). Approval was granted in 1996, in part based on the results of the two-part National Institute of Neurological Disorders and Stroke (NINDS) rtPA Stroke Trial. A large, randomized, double-blind, placebo-controlled study by the NINDS indicated a beneficial effect of rtPA (Group TNIoNDaSr-PSS 1995). It was shown that treatment with intravenous rtPA within 3 h of ischemic stroke onset improved the clinical outcome at 3 months, despite an increased incidence of symptomatic intracerebral hemorrhage. In a 12-month follow-up study, acute ischemic stroke patients who were treated with rtPA within 3 h after the symptom onset were at least 30% more likely to have minimal or no disability at 12 months than placebo-treated patients, indicating a sustained benefit of rtPA for such patients (Kwiatkowski et al. 1999).

Despite the wide acceptance of rtPA in ischemic stroke treatment, there still remains controversy. Besides a risk of intracranial hemorrhage (Group TNIoNDaSr-PSS 1995; Kwiatkowski et al. 1999; Adams et al. 2003), there are also other potential adverse experiences including systemic bleeding, myocardial rupture, and allergic reactions including anaphylaxis (Becker et al. 1999). Therefore, intravenous administration of rtPA is only recommended for carefully selected patients who are eligible according to the treatment criteria and who can receive the medication within 3 h of onset of the stroke (Adams et al. 2003). The restrictive 3-h time window necessary for the use of rtPA further limits the number of patients who will receive the intervention. It is estimated that only about 1–2% of patients with acute ischemic stroke have received this intervention (Hacke et al. 1999). Subsequent to the NINDS studies, five clinical trials have tested the use of intravenous rtPA up to 6 h after the onset of the stroke. Up to date, there still remain contradictory actions regarding later administration of intravenous rtPA; the European Medicines Agency expands approval of intravenous rtPA to the 3- to 4.5-h window but the US FDA declines to do so (Jauch et al. 2013). Yet, based on clinical trial evidence, it is still fundamentally important to minimize the total ischemic time and to restore the blood flow to the threatened but not yet infarcted tissue as soon as feasible (Khatri et al. 2009). Indeed, health systems are highly recommended to set a goal: increase the percentage of door-to-needle time of 60 min, i.e., stroke patients treated within 60 min of presentation to at least 80% (Jauch et al. 2013).

Due to the narrow time window and adverse effects after an ischemic stroke such as hemorrhagic transformation, ischemia/reperfusion damage and reocclusion after recanalization, neuroscientists have become more emphasized on neuroprotection in the treatment of ischemic stroke. The idea behind neuroprotection is different from reperfusion; its aim is to directly affect the brain tissue in order to salvage cells in the still-viable penumbra area. Currently, the most studied pharmacological neuroprotective agents include glutamate receptor antagonists, ion channel modulators, anti-inflammatory agents, and free radical scavengers (Cheng et al. 2004; Jauch et al. 2013); simply speaking, agents that target almost every single component of the ischemia cascade. There have been more than 1000 published reports of various experimental neuroprotective treatments for acute stroke and over 100 clinical trials (Kidwell et al. 2001; O'Collins et al. 2006; Jauch et al. 2013). Despite the determined efforts in basic research and clinical investigation, neuroprotective therapies for acute ischemic stroke remain unsuccessful. They all appear to be effective in initial studies using a variety of animal stroke models; however, the outcomes of the subsequent clinical trials are far from satisfactory. Most clinical trials testing these therapies have produced disappointing results and none of them has been proven effective in clinical practice. Unfortunately, worse outcomes are observed in treated patients in some circumstances (Jauch et al. 2013).

9.4 Stroke and Traditional Medicine

With an increasing population of elderly people worldwide, stroke has become a major healthcare problem; yet, advances are being made. The incidence of stroke has been reduced by preventive measures such as controlling hypertension, hypercholesterolemia, and smoking. Besides staying vigilant and committed to improving overall stroke care in the public sector and medical field, there remains an urgent need for effective and widely applicable pharmacological treatment. This lack of treatment for stroke may explain the growing attention in traditional medicine in the last few decades, which have accumulated extensive observational and anecdotal experiences over the past millenniums.

In 2003, the World Health Organization (WHO) defined traditional medicine as "health practices, approaches, knowledge, and beliefs incorporating plant, animal and mineral based medicines, spiritual therapies, manual techniques and exercises, applied singularly or in combination to treat, diagnose, and prevent illnesses or maintain well-being" (World Health Organization 2003). Traditional medicine, also termed as alternative medicine, arises in ancient times before the establishment of modern Western medicine in various regions around the world over generations. It is receiving increasing popularity in the Western society. In fact, there are approximately 80 % of the population who depend on traditional medicine for primary health care in some African and Asian countries, while in developed regions such as Europe and North America, 70–80 % of the population has at least once been treated by traditional medicine (World Health Organization 2003).

The use of traditional medicine in China, also known as traditional Chinese medicine (TCM), is very prevalent due to not only its long history but also its recognized efficacy. Chinese herbal medicine has been used to treat ischemic stroke for ages. Dozens of Chinese herbs, either crude or extract, have been tested in various preclinical animal studies and clinical trials. Some of the actions of TCM in stroke therapy can be explained based on the pathophysiology of cerebral ischemia. They contain agents with antioxidative, anti-inflammatory, and antithrombotic properties. For example, many herbs contain flavonoids, whose antioxidative activity has been shown in vitro (Bagchi et al. 1999). Flavonoids are also shown to have anti-inflammatory activities (Howes et al. 2003; Yamamoto and Gaynor

2001), which may be desirable in stroke treatment. In addition, some herbs can dilate blood vessels and suppress platelet aggregation (Bei et al. 2007).

9.5   Herbs Used for Treatment or Prevention of Stroke in Traditional Chinese Medicine

There are more than 100 herbs that have been used in stroke prevention or treatment. Some of their active ingredients have been described in detail (Zhou and Xiao 1997; Gong and Sucher 1999; Kim 2005). Examples of the more commonly used herbs are listed below in alphabetical order. Among them, *Lycium barbarum* has received much interest.

Acanthropanax senticosus harms
Angelica sinensis, A. gigas
Astragalus membranaceus
Bombycis corpus
Carthamus tinctorius
Corydalis yanhusuo
Ginkgo biloba
Ligusticum wallichii Franchat
Magnolia officinalis Rehder et Wilson
Paeonia suffruticosa Andrews, *Paeonia lactiflora* Pall
Panax ginseng
Pueraria lobata
Rhodiola rosea L., Rhodiola sacra S. H. FU, *Rhodiola sachalinensis* A. BOR
Salvia miltiorrhiza bunge
Schisandra chinensis
Scutellaria baicalensis
Sophora japonica L.
Stephania tetrandra S. Moore

9.5.1   *Stroke and Lycium Barbarum*

L. barbarum (also known as Goji, Gouqizi, Wolfberry, *Fructus Lycii*) is a deciduous woody plant growing natively in Asia and southeastern Europe, producing light purple flower and orange-red berry fruit. The fruit of *L. barbarum* has been widely used in Chinese herbal recipes for millenniums. In TCM literature such as 神龍本草經 and 本草綱目, *L. barbarum* can nourish the liver and kidney as well promoting health and longevity. *L. barbarum* has an attractive red color, leading to the belief that it has a role in strengthening eyesight and protecting the eyes.

There are abundant ingredients in the fruit of *L. barbarum* including β-carotene, thiamine, riboflavin, betaine, zeaxanthin, electrolytes, trace mineral, amino acids, vitamins, lipid, and polysaccharides. In *L. barbarum*, the polysaccharides constitute more than 40% of the fruit extract (Chang and So 2008). Numerous experimental studies have revealed that *L. barbarum* has a wide array of functions, which may be due to its high polysaccharide content. These include antiaging, antitumor, cytoprotective, neuromodulation, and immune modulation effects (Chang and So 2008). In addition, a neuroprotective role of extracts of *L. barbarum* containing mostly *L. barbarum* polysaccharides (LBP) in the nervous system has been strongly indicated by in vitro and in vivo studies.

LBP pretreatment can protect the cultured primary cortical neurons from β-amyloid peptide neurotoxicity (Yu et al. 2005). It further protects neurons against β-amyloid-induced apoptosis by reducing the activity of both caspase-3 and caspase-2 (Yu et al. 2007). In another study where primary cultured rat hippocampal neurons were exposed to oxygen-glucose deprivation, LBP treatment attenuated neuronal damage and dose dependently inhibited lactose dehydrogenase release (Rui et al. 2012). The possible mechanisms of LBP-mediated mechanisms in these in vitro studies include the reduction in the phosphorylation of JNK-1 (Thr183/Tyr185) and its substrates c-Jun-I (Ser 73) and c-Jun-II (Ser 63) (Yu et al. 2005), reduction in the phosphorylation of PKR triggered by β-amyloid peptide(Yu et al. 2007), enhancement of activities of superoxide dismutase (SOD) and glutathione peroxidase (GSH-Px), decreased malondialdehyde (MDA) content, as well as reduction in mitochondrial membrane potential (MMP) (Rui et al. 2012).

More importantly, oral LBP pretreatment is beneficial in animal models of experimental stroke. In patients with ischemic stroke, approximately 80% thrombotic or embolic strokes occur in the territory of the middle cerebral artery (MCA) (Rousselet et al. 2012). To mimic focal brain damages similar to those occurring in human stroke, the intraluminal method to induce middle cerebral artery occlusion (MCAO) was first developed in rats by Koizumi et al. in 1986 (Koizumi et al. 1986) and was then modified to be performed in mice due to the advancement in murine genetic manipulation technology (Chan et al. 1993; Huang et al. 1994; Yang et al. 1994). Here, a monofilament is advanced through the external carotid artery into the internal carotid artery until it occludes the MCA at its origin. Blood flow is restored when the filament is removed and reperfusion occurs (Lo et al. 2005, 2007; Yeung et al. 2010; Li et al. 2012; Yang et al. 2012). This is a highly reproducible and widely accepted model for cerebral ischemia/reperfusion (I/R) injury and is relevant for stroke studies. In studies using such animal models, LBP effectively improves neurological deficits, decreased infarct size and cerebral edema after MCAO-induced I/R injury, indicating its neuroprotective effects (Yang et al. 2012). Moreover, blood-brain barrier disruption, aquaporin-4 upregulation and glial activation are attenuated by LBP pretreatment (Yang et al. 2012). The antioxidative (Wang et al. 2013) and antiapoptotic (Wang et al. 2014) effects of LBP have also been suggested to be involved in its neuroprotective mechanisms. LBP could decrease MDA content and increase SOD and GSH-Px activities in the ischemic brain (Wang et al. 2013). Moreover, LBP suppressed the upregulation of Bax and caspase-3 while inhibiting the reduction of

Bcl-2, events that are induced by focal cerebral ischemic injury (Wang et al. 2014). These results suggest that LBP may be used as a prophylactic neuroprotectant in patients at high risk for ischemic stroke.

LBP may also have neuroprotective effects in the retina after I/R injury. Retinal I/R injury is common in many ocular diseases such as amaurosis fugax, glaucoma, and diabetic retinopathy. In a murine retinal I/R injury model, LBP pretreatment could effectively protect the retina from neuronal death, glial activation, and oxidative stress (Li et al. 2011). After I/R injury in mouse retina, there was apoptosis and decreased viable cell count in the ganglion cell layer and the inner nuclear layer of the vehicle-treated I/R retina. In addition, increased retinal thickness, glial fibrillary acidic protein (GFAP) activation, aquaporin 4 (AQP4) upregulation, immunoglobulin G (IgG) extravasations, and proteinase-activated receptor (PAR) expression level were observed. These phenotypic changes could almost be completely reversed by LBP pretreatment (Li et al. 2011). LBP was also shown to provide neuroprotection by downregulating the receptor for advanced glycation end-products (RAGE), endothelin 1 (ET-1), β-amyloid and advanced glycation end-products (AGE) in the retina in a murine model of ocular hypertension (Mi et al. 2012). A smaller loss of retinal ganglion cells, higher level of occludin protein, and recovery of the blood vessel density were observed. In a rat ocular hypertension model, LBP pretreatment protected the retinal ganglion cells (Chan et al. 1993). Later studies reported the ability of LBP pretreatment to increase Nrf2 nuclear accumulation and heme oxygenase 1 (HO-1) expression in the retina using a similar retinal I/R injury model (He et al. 2014).

Conclusion

Stroke is a devastating disease and causes major social burden worldwide. Despite the vast diversity of investigations using various in vitro and in vivo experimental models as well as clinical trials, there is still a lack of effective and widely applicable pharmacological treatments for ischemic stroke patients to salvage neuronal death. The search continues while the public waits and yearns for better measures. Many people have thus turned to the use of TCM that has extensive observational and anecdotal experience inherited and accumulated over thousands of years. Its safety and efficacy, although not proven according to the present standards, should not be doubted. Studies on LBP so far have strongly supported a beneficial neuroprotective role of LBP pretreatment. LBP pretreatment with continuous daily supplementation protected the brain and retina, both functionally and morphologically, from I/R injury in experimental models of ischemic stroke and ischemic retinopathy. This regimen holds great promise in serving as a prophylactic neuroprotectant in patients at high risk for ischemic stroke as well as reducing irreversible neuronal death in ischemic retinopathies. The protective effects are thought to be associated with LBP's antiapoptotic, antioxidative, and anti-inflammatory properties. Similar to other TCM, LBP is a mixture of various components including glucose, arabinose, galacturonic acid, and galactose (Yu et al. 2007). It will be of great interest to investigate and delineate the active ingredients of LBP as well as to compare their isolated effects in ischemic condition with the synergistic effects subsequently. Yet, as with other neuroprotective strategy for ischemic stroke, further scientific research together with large-scale, properly designed randomized controlled trials followed by appropriate meta-analysis are necessary and essential before such treatments can be recommended for general clinical use.

References

Adams HP Jr, Adams RJ, Brott T, del Zoppo GJ, Furlan A, Goldstein LB, Grubb RL, Higashida R, Kidwell C, Kwiatkowski TG, Marler JR, Hademenos GJ. Guidelines for the early management of patients with ischemic stroke: a scientific statement from the stroke council of the American stroke association. Stroke. 2003;34(4):1056–83. doi:10.1161/01.STR.0000064841.47697.22.

Astrup J, Siesjo BK, Symon L. Thresholds in cerebral ischemia—the ischemic penumbra. Stroke. 1981;12(6):723–5.

Bagchi M, Milnes M, Williams C, Balmoori J, Ye XM, Stohs S, Bagchi D. Acute and chronic stress-induced oxidative gastrointestinal injury in rats, and the protective ability of a novel grape seed proanthocyanidin extract. Nutr Res. 1999;19(8):1189–99.

Barber PA, Auer RN, Buchan AM, Sutherland GR. Understanding and managing ischemic stroke. Can J Physiol Pharmacol. 2001;79(3):283–96.

Barone FC, Tuma RF, Legos JJ, Erhardt JA, Parsons A. Brain inflammation, cytokines, and p38 MAP kinase signaling in stroke. In: Lin RCS, editor. New concepts in cerebral ischemia. Boca Raton: CRC; 2002.

Becker RC, Hochman JS, Cannon CP, Spencer FA, Ball SP, Rizzo MJ, Antman EM. Fatal cardiac rupture among patients treated with thrombolytic agents and adjunctive thrombin antagonists: observations from the thrombolysis and thrombin inhibition in myocardial infarction 9 study. J Am Coll Cardiol. 1999;33(2):479–87.

Bei W, Peng W, Zang L, Xie Z, Hu D, Xu A. Neuroprotective effects of a standardized extract of diospyros kaki leaves on MCAO transient focal cerebral ischemic rats and cultured neurons injured by glutamate or hypoxia. Planta Medica. 2007;73(7):636–43. doi:10.1055/s-2007-981532.

Chan PH, Kamii H, Yang G, Gafni J, Epstein CJ, Carlson E, Reola L. Brain infarction is not reduced in SOD-1 transgenic mice after a permanent focal cerebral ischemia. Neuroreport. 1993;5(3):293–6.

Chang RC, So KF. Use of anti-aging herbal medicine, *Lycium barbarum*, against aging-associated diseases. What do we know so far? Cell Mol Neurobiol. 2008;28(5):643–52. doi:10.1007/s10571-007-9181-x.

Cheng YD, Al-Khoury L, Zivin JA. Neuroprotection for ischemic stroke: two decades of success and failure. NeuroRx. 2004;1(1):36–45. doi:10.1602/neurorx.1.1.36.

Dirnagl U, Iadecola C, Moskowitz MA. Pathobiology of ischaemic stroke: an integrated view. Trends Neurosci. 1999;22(9):391–7.

Doyle KP, Simon RP, Stenzel-Poore MP. Mechanisms of ischemic brain damage. Neuropharmacology. 2008;55(3):310–8. doi:10.1016/j.neuropharm.2008.01.005.

Goldstein LB, Adams R, Alberts MJ, Appel LJ, Brass LM, Bushnell CD, Culebras A, Degraba TJ, Gorelick PB, Guyton JR, Hart RG, Howard G, Kelly-Hayes M, Nixon JV, Sacco RL. Primary prevention of ischemic stroke: a guideline from the American heart association/American stroke association stroke council: cosponsored by the atherosclerotic peripheral vascular disease interdisciplinary working group; cardiovascular nursing council; clinical cardiology council; nutrition, physical activity, and metabolism council; and the quality of care and outcomes research interdisciplinary working group: the American academy of neurology affirms the value of this guideline. Stroke. 2006;37(6):1583–633. doi:10.1161/01.STR.0000223048.70103.F1.

Gong X, Sucher NJ. Stroke therapy in traditional Chinese medicine (TCM): prospects for drug discovery and development. Trends Pharmacol Sci. 1999;20(5):191–6.

Group TNIoNDaSr-PSS. Tissue plasminogen activator for acute ischemic stroke. N Engl J Med. 1995;333(24):1581–7. doi:10.1056/NEJM199512143332401.

Hacke W, Brott T, Caplan L, Meier D, Fieschi C, von Kummer R, Donnan G, Heiss WD, Wahlgren NG, Spranger M, Boysen G, Marler JR. Thrombolysis in acute ischemic stroke: controlled trials and clinical experience. Neurology. 1999;53(7 Suppl. 4):S3–14.

He M, Pan H, Chang RC, So KF, Brecha NC, Pu M. Activation of the Nrf2/HO-1 antioxidant pathway contributes to the protective effects of *Lycium barbarum* polysaccharides in the rodent retina after ischemia-reperfusion-induced damage. PloS ONE. 2014;9(1):e84800. doi:10.1371/journal.pone.0084800.

Howes MJ, Perry NS, Houghton PJ. Plants with traditional uses and activities, relevant to the management of Alzheimer's disease and other cognitive disorders. Phytother Res. 2003;17(1):1–18. doi:10.1002/ptr.1280.

Huang Z, Huang PL, Panahian N, Dalkara T, Fishman MC, Moskowitz MA. Effects of cerebral ischemia in mice deficient in neuronal nitric oxide synthase. Science. 1994;265(5180):1883–5.

Jauch EC, Saver JL, Adams HP Jr, Bruno A, Connors JJ, Demaerschalk BM, Khatri P, McMullan PW Jr, Qureshi AI, Rosenfield K, Scott PA, Summers DR, Wang DZ, Wintermark M, Yonas H, American Heart Association Stroke C, Council on Cardiovascular N, Council on Peripheral Vascular D, Council on Clinical C. Guidelines for the early management of patients with acute ischemic stroke: a guideline for healthcare professionals from the American Heart Association/American Stroke Association. Stroke. 2013;44(3):870–947. doi:10.1161/STR.0b013e318284056a.

Khatri P, Abruzzo T, Yeatts SD, Nichols C, Broderick JP, Tomsick TA, Ims I, Investigators II. Good clinical outcome after ischemic stroke with successful revascularization is time-dependent. Neurology. 2009;73(13):1066–72. doi:10.1212/WNL.0b013e3181b9c847.

Kidwell CS, Liebeskind DS, Starkman S, Saver JL. Trends in acute ischemic stroke trials through the 20th century. Stroke. 2001;32(6):1349–59.

Kim H. Neuroprotective herbs for stroke therapy in traditional eastern medicine. Neurol Res. 2005;27(3):287–301. doi:10.1179/016164105X25234.

Koizumi J, Yoshida Y, Nakazawa T, Ooneda G. Experimental studies of ischemic brain edema, I: a new experimental model of cerebral embolism in rats in which recirculation can be introduced in the ischemic area. Jpn J Stroke. 1986;8:1–8

Kwiatkowski TG, Libman RB, Frankel M, Tilley BC, Morgenstern LB, Lu M, Broderick JP, Lewandowski CA, Marler JR, Levine SR, Brott T. Effects of tissue plasminogen activator for acute ischemic stroke at one year. National institute of neurological disorders and stroke recombinant tissue plasminogen activator stroke study group. N Engl J Med. 1999;340(23):1781–7. doi:10.1056/NEJM199906103402302.

Li SY, Yang D, Yeung CM, Yu WY, Chang RC, So KF, Wong D, Lo AC. *Lycium barbarum* polysaccharides reduce neuronal damage, blood-retinal barrier disruption and oxidative stress in retinal ischemia/reperfusion injury. PloS ONE. 2011;6(1):e16380. doi:10.1371/journal.pone.0016380.

Li SY, Yang D, Fu ZJ, Woo T, Wong D, Lo AC. Lutein enhances survival and reduces neuronal damage in a mouse model of ischemic stroke. Neurobiol Dis. 2012;45(1):624–32. doi:10.1016/j.nbd.2011.10.008.

Lin RCS. New concepts in cerebral ischemia. Boca Raton: CRC; 2002.

Lo EH, Dalkara T, Moskowitz MA. Mechanisms, challenges and opportunities in stroke. Nat Rev Neurosci. 2003;4(5):399–415. doi:10.1038/nrn1106.

Lo AC, Chen AY, Hung VK, Yaw LP, Fung MK, Ho MC, Tsang MC, Chung SS, Chung SK. Endothelin-1 overexpression leads to further water accumulation and brain edema after middle cerebral artery occlusion via aquaporin 4 expression in astrocytic end-feet. J Cereb Blood Flow Metab. 2005;25(8):998–1011. doi:10.1038/sj.jcbfm.9600108.

Lo AC, Cheung AK, Hung VK, Yeung CM, He QY, Chiu JF, Chung SS, Chung SK. Deletion of aldose reductase leads to protection against cerebral ischemic injury. J Cereb Blood Flow Metab. 2007;27(8):1496–509. doi:10.1038/sj.jcbfm.9600452.

Love S. Oxidative stress in brain ischemia. Brain Pathol. 1999;9(1):119–31

Mi XS, Feng Q, Lo AC, Chang RC, Lin B, Chung SK, So KF. Protection of retinal ganglion cells and retinal vasculature by *Lycium barbarum* polysaccharides in a mouse model of acute ocular hypertension. PloS ONE. 2012;7(10):e45469. doi:10.1371/journal.pone.0045469.

O'Collins VE, Macleod MR, Donnan GA, Horky LL, van der Worp BH, Howells DW. 1026 experimental treatments in acute stroke. Ann Neuro. 2006;59(3):467–77. doi:10.1002/ana.20741.

Olsen TS, Larsen B, Herning M, Skriver EB, Lassen NA. Blood flow and vascular reactivity in collaterally perfused brain tissue. Evidence of an ischemic penumbra in patients with acute stroke. Stroke. 1983;14(3):332–41.

Rousselet E, Kriz J, Seidah NG. Mouse model of intraluminal MCAO: cerebral infarct evaluation by cresyl violet staining. J Vis Exp. 2012;69:4038. doi:10.3791/4038.

Rui C, Yuxiang L, Yinju H, Qingluan Z, Yang W, Qipeng Z, Hao W, Lin M, Juan L, Chengjun Z, Yuanxu J, Yanrong W, Xiuying D, Wannian Z, Tao S, Jianqiang Y. Protective effects of *Lycium barbarum* polysaccharide on neonatal rat primary cultured hippocampal neurons injured by oxygen-glucose deprivation and reperfusion. J Mol Histol. 2012;43(5):535–42. doi:10.1007/s10735-012-9420-4.

Wang HB, Li YX, Hao YJ, Wang TF, Lei Z, Wu Y, Zhao QP, Ang H, Ma L, Liu J, Zhao CJ, Jiang YX, Wang YR, Dai XY, Zhang WN, Sun T, Yu JQ. Neuroprotective effects of LBP on brain ischemic reperfusion neurodegeneration. Euro Rev Med Pharmacol Sci. 2013;17(20):2760–5.

Wang T, Li Y, Wang Y, Zhou R, Ma L, Hao Y, Jin S, Du J, Zhao C, Sun T, Yu J. *Lycium barbarum* polysaccharide prevents focal cerebral ischemic injury by inhibiting neuronal apoptosis in mice. PloS ONE. 2014;9(3):e90780. doi:10.1371/journal.pone.0090780.

Wityk RJ, Stern BJ. Ischemic stroke: today and tomorrow. Crit Care Med. 1994;22(8):1278–93

World Health Organization . World Health Organization fact sheet No. 134, Revised May 2003 ed. Geneva: World Health Organization; 2003.

Yamamoto Y, Gaynor RB. Therapeutic potential of inhibition of the NF-kappaB pathway in the treatment of inflammation and cancer. J Clin Invest. 2001;107(2):135–42. doi:10.1172/JCI11914.

Yang G, Chan PH, Chen J, Carlson E, Chen SF, Weinstein P, Epstein CJ, Kamii H. Human copper-zinc superoxide dismutase transgenic mice are highly resistant to reperfusion injury after focal cerebral ischemia. Stroke.1994;25(1):165–70.

Yang D, Li SY, Yeung CM, Chang RC, So KF, Wong D, Lo AC. *Lycium barbarum* extracts protect the brain from blood-brain barrier disruption and cerebral edema in experimental stroke. PloS ONE. 2012;7(3):e33596. doi:10.1371/journal.pone.0033596.

Yeung CM, Lo AC, Cheung AK, Chung SS, Wong D, Chung SK. More severe type 2 diabetes-associated ischemic stroke injury is alleviated in aldose reductase-deficient mice. J Neurosci Res. 2010;88(9):2026–34. doi:10.1002/jnr.22349.

Yu MS, Leung SK, Lai SW, Che CM, Zee SY, So KF, Yuen WH, Chang RC. Neuroprotective effects of anti-aging oriental medicine *Lycium barbarum* against beta-amyloid peptide neurotoxicity. Exp Gerontol. 2005;40(8–9):716–27. doi:10.1016/j.exger.2005.06.010.

Yu MS, Lai CS, Ho YS, Zee SY, So KF, Yuen WH, Chang RC. Characterization of the effects of anti-aging medicine Fructus lycii on beta-amyloid peptide neurotoxicity. Int J Mol Med. 2007;20(2):261–8.

Zhou S, Xiao P. A modern practical handbook of neurology and psychosis of the integration of traditional Chinese and Western medicine. Hunan: Hunan Science and Technology; 1997.

Chapter 10
Secondary Degeneration After Partial Optic Nerve Injury and Possible Neuroprotective Effects of *Lycium Barbarum* (Wolfberry)

Hong-Ying Li, Henry HL Chan, Patrick HW Chu, Raymond Chuen-Chung Chang and Kwok-Fai So

Abstract Secondary degeneration occurs commonly in a range of neurodegenerative diseases, including glaucoma. Partial optic nerve transection (PONT) model was established in the last decade and was good for studying secondary degeneration in retinas and optic nerves. The results from the published papers about PONT showed that the mechanisms—apoptosis, necrosis, autophagy, oxidative stress, calcium overload, mitochondria, activation of c-jun, water channel change, and glial cells (microglia, astrocytes and oligodendrocytes)—were involved in secondary degeneration after PONT. In addition to the cell bodies and the axons of retinal ganglion cells (RGCs), other cells in the layers outside the ganglion cell layer were also affected according to the measurement of multifocal electroretinogram (mfERG) by our group. *Lycium barbarum* (*L. barbarum*) is a traditional medicine in the oriental world and has long been used as a functional food and for medicinal purposes. The data from our group and others showed that the polysaccharides extracted from

K.-F. So (✉) · H.-Y. Li · R. C.-C. Chang
GHM Institute of CNS Regeneration and Guangdong Key Laboratory of Brain Function and Diseases, Jinan University, Guangzhou, People's Republic of China
e-mail: hrmaskf@hku.hk

H.-Y. Li
Department of Anatomy, Jinan University School of Medicine, Guangzhou, People's Republic of China

H.-Y. Li · K.-F. So
Department of Ophthamology, The University of Hong Kong, Hong Kong, People's Republic of China

H.-Y. Li · R. C.-C. Chang · K.-F. So
The State Key Laboratory of Brain and Cognitive Science and the Research Centre of Heart, Brain, Hormone and Healthy Aging, The University of Hong Kong, Hong Kong, People's Republic of China

R. C.-C. Chang · K.-F. So
Department of Anatomy, The University of Hong Kong, Hong Kong, People's Republic of China

H. H. Chan · P. H. Chu
Laboratory of Experimental Optometry (Neuroscience), School of Optometry, The Hong Kong Polytechnic University, Hong Kong, People's Republic of China

© Springer Science+Business Media Dordrecht 2015
R. C-C. Chang, K-F. So (eds.), *Lycium Barbarum and Human Health*,
DOI 10.1007/978-94-017-9658-3_10

135

L. barbarum (LBP) were neuroprotective in different animal models, including the PONT model. Our results showed that LBP could inhibit secondary degeneration of the cell bodies of RGCs rather than primary degeneration as well as preserve the function of retinas measured by mfERG. These effects are related with the antioxidant function of LBP, inhibition of c-jun N-terminal kinase (JNK) pathway in the retinas after PONT. Other possible mechanisms involved in LBP's neuroprotective effects for secondary degeneration are immunomodulatory effects, preservation of synapses, and modulation of autophagy.

Keywords Optic nerve · Secondary degeneration · *Lycium barbarum* · Neuroprotection

10.1   Introduction

After traumatic damages and during acute and chronic diseases in the central nervous system (CNS), such as glaucoma, secondary degeneration universally occurs. A great number of mechanisms are associated with secondary degeneration including apoptosis, necrosis, autophagy, oxidative stress, excitotoxicity, derangements in ionic homeostasis, and calcium influx. Glial cells (microglia, astrocytes, and oligodendrocytes) also play a role in secondary injury. Partial optic nerve (ON) transection (PONT), established in the last decade, is a suitable model for glaucoma study. Compared with the complete optic nerve transection (CONT) model and ON crush model, the merit of this model is the feasibility to separate primary degeneration from secondary degeneration in location. Therefore, it is a good tool for studying secondary degeneration. The first part of this chapter will focus on the research progress about the mechanisms of secondary degeneration using the PONT model.

In Chinese medicine, *Lycium barbarum* (*L. barbarum*) has been used as an "upper-class herb" for many years, and it is good for the vision sight, nourishing the "kidney" and protecting the "liver" (Junlin and Aicheng 2002). *L. barbarum* is made up of many components, for example, polysaccharides (Chang and So 2008; Ho et al. 2009), taurine (Song et al. 2011), betaine (Xie et al. 2001; Lee et al. 2004), zeaxanthin (Kim et al. 2002), beta-carotene (Inbaraj et al. 2008), beta-sitosterol (Xie et al. 2001), flavonoids (Lee et al. 2004), vitamins (Zhang et al. 2011), amino acids (Wang et al. 2012), fatty acids (Honglin et al. 2009), and trace minerals (Yang et al. 2012b). The neuroprotective effects of *L. barbarum* have been shown in both cell culture models and animal models. *L. barbarum* has also been studied extensively in eye diseases including diabetic retinopathy, ischemia model, and glaucoma. The protective mechanisms of *L. barbarum* are achieved through the properties including antioxidation (Li 2007; Cheng and Kong 2011; Li et al. 2011; Shan et al. 2011; Xiao et al. 2012), anti-excitotoxicity (Ho et al. 2009), anti-inflammation (Wu et al. 2011; Xiao et al. 2012), and anti-apoptosis (Ho et al. 2010; Li et al. 2011; Song et al. 2012). The second part of this chapter will give details about the neuroprotective effects of *L. barbarum* in secondary degeneration in eye diseases and summarize the possible mechanisms.

10.2   Secondary Degeneration

10.2.1   Secondary Degeneration in Glaucoma

The degeneration of neurons and glial cells which is caused by primary pathological events and occurs early in the pathological process is called primary degeneration. However, the neurons and glial cells which are not or only partially affected by the primary damage will also die, and this is called secondary degeneration. Secondary degeneration occurs universally in the CNS after traumatic injuries and in acute diseases and chronic neurodegenerative diseases. For example, secondary degeneration emerged after brain trauma (Stoica and Faden 2010), spinal cord injury (Hausmann 2003; Oyinbo 2011), stroke (Guimaraes et al. 2009), and also in chronic neurodegenerative diseases, such as Alzheimer's disease, Parkinson's disease, and amyotrophic lateral sclerosis (Stewart and Appel 1988).

Glaucoma is a chronic, progressive neurodegenerative disease of the CNS, which is limited to the visual system. Decreasing the ocular hypertension (OH) with surgery has been shown to be the only confirmed effective therapy for glaucoma. But the situation of some patients deteriorates after the OH has been controlled. Based on this fact, secondary degeneration of retinal ganglion cells (RGCs) should exist in glaucoma (Tezel 2006; Nickells 2007; Tezel 2008). Mechanisms such as deprivation of neurotrophic factor, failure in axonal transport, apoptosis, excitotoxicity, oxidative stress, dysfunctions of glial cells, and loss of synaptic connectivity are involved in RGC death in glaucoma although it is unknown which ones are specifically related with primary or secondary degeneration. This part has been well summarized by Almasieh et al. (2012). Apart from RGCs, amacrine cells and photoreceptors are also affected in glaucoma by transsynaptic secondary degeneration (Calkins 2012). Prevention or delay of secondary degeneration presents a new direction for the treatment of glaucoma and other optic neuropathies.

10.2.2   Secondary Degeneration After ON Injuries

As mentioned above, the prevention of secondary degeneration was a promising direction for the therapy of glaucoma. The axonal degeneration was precedent to the death of RGC bodies in glaucoma (Calkins 2012). Therefore, ON injury models were widely used in the study of glaucoma. Three kinds of ON injury models have been used for studying the mechanisms of RGC degeneration and the possible therapeutic measures: the CONT model, the ON crush model, and the PONT model. Although it seems that some RGCs will take a longer time period to die than others after CONT, the occurrence of secondary degeneration after CONT is uncertain because all the axons are transected, and the degeneration of axons can lead to the degeneration of RGC bodies. After partial crush of ON, some RGCs survive longer than others. However, the number of axons influenced was unknown and the intact

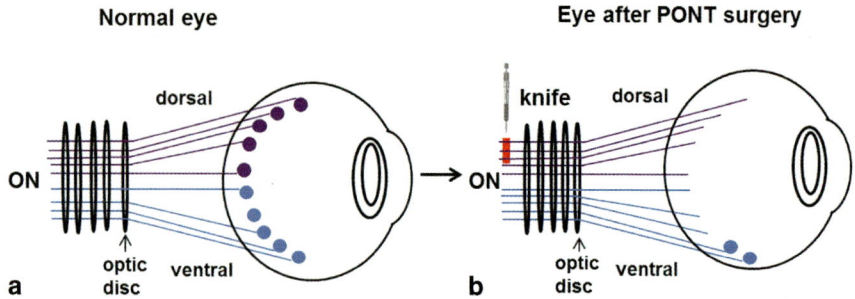

Fig. 10.1 The schematic diagram to show the PONT model. The *lines* represent the axons in the ON and the *circles* indicate the cell bodies of RGCs: the *purple* structures locate in the dorsal parts of the retina and ON and the *blue* structures locate in the ventral parts of the retina and ON. The *red* line shows the cut site after PONT in the dorsal ON. **a** The cell bodies and axons of RGCs are intact in normal eye and connect with each other. **b** After PONT surgery, in addition to the degeneration of the cell bodies of RGCs whose axons are transected, the cell bodies whose axons are reserved after PONT also degenerate

and damaged axons were mingled together. Therefore, it was nearly impossible to distinguish the cells which required a longer time for degeneration from others in location (Yoles and Schwartz 1998). In order to clarify this issue, a PONT model has been established using monkeys in 2001 by Levkovitch-Verbin et al. In this model, only the dorsal part of ON was damaged, but cell body degeneration of RGCs whose axons were kept intact after injury occurred in the inferior retinas (Levkovitch-Verbin et al. 2001). Wistar rats were used to repeat this model by the same group, and a similar result was obtained in rat PONT model (Levkovitch-Verbin et al. 2003). Convincing information suggested that some of the neighboring RGCs would die from the influence of biochemical events that derived from the original damage in the PONT model. What's more important, the location of the RGC bodies' secondary degeneration was confirmed to the inferior retinas after PONT in Wistar rats. Therefore, the PONT model provides a good tool for the study of secondary degeneration. The PONT model is described in Fig. 10.1. This model has been used extensively for studying the mechanisms of secondary degeneration and the possible neuroprotective agents from its establishment (Yoles and Schwartz 1998; Levkovitch-Verbin et al. 2003; Fitzgerald et al. 2009a, b, 2010a, b; Levkovitch-Verbin et al. 2010; Selt et al. 2010; Levkovitch-Verbin et al. 2011; Payne et al. 2011, 2012; Wells et al. 2012; Chu et al. 2013; Cummins et al. 2013; Fitzgerald et al. 2013; Li et al. 2013; Payne et al. 2013; Savigni et al. 2013; Szymanski et al. 2013).

10.2.3 *Mechanisms of Secondary Degeneration After PONT*

The PONT model has been conducted in three species of rats, including Wistar rats (Levkovitch-Verbin et al. 2003), PVG hooded rats (Fitzgerald et al. 2009a), and Sprague Dawley (SD) rats (Li et al. 2013). After partial injury in the dorsal ON (about one quarter in Wistar rats and SD rats or one third in PVG hooded rats), the axons in the central and the ventral ON would be vulnerable to secondary degeneration (Levkovitch-Verbin et al. 2003; Fitzgerald et al. 2009b, 2010b; Li et al. 2013). In the retinas, the localization of primary and secondary degeneration of the RGC bodies should be based on the topography of the transected axons in the three different strains (Levkovitch-Verbin et al. 2003; Fitzgerald et al. 2009b; Li et al. 2013). The mechanisms underlying both primary and secondary degeneration after PONT were studied simultaneously in some experiments. But the scope of this chapter is mainly limited to the mechanisms of secondary degeneration after PONT. Several mechanisms were assumed to be related with secondary degeneration after PONT, including:

10.2.3.1   Apoptosis, Necrosis, and Autophagy

In the retinas, both apoptosis and necrosis contributed to secondary degeneration of the RGC bodies after PONT (Levkovitch-Verbin et al. 2010; Fitzgerald et al. 2009b). In the ON, mitochondrial autophagic profiles were observed in the areas vulnerable to secondary degeneration (Cummins et al. 2013).

10.2.3.2   Oxidative Stress

Oxidative stress, indicated by increased expression of manganese superoxide dismutase (MnSOD) and decreased catalase activity, occurred as early as 5 min after injury in the ventral ON (area vulnerable to secondary degeneration; Fitzgerald et al. 2010a; Wells et al. 2012). In the retinas, the oxidative stress indicated by increasing expression of MnSOD began at 24 h after PONT in secondary degeneration both in PVG hooded rats and SD rats (Fitzgerald et al. 2010a; Li et al. 2013). Therefore, oxidative stress was involved in the secondary degeneration of both axons and RGC bodies after PONT.

10.2.3.3   Calcium Overload

The result after PONT indicated a redistribution of calcium from internal stores to the cytosol (Wells et al. 2012). It is known that calcium accumulation can lead to the overproduction of reactive oxygen species (ROS) in mitochondria (Lewen et al. 2000; Begemann et al. 2010). In addition, the immunoreactivity of glutamate receptor subunit GluR1, a subunit of functional α-amino-3-hydroxy-5-methyl-4-isoxazolepropionic acid (AMPA) receptors, was increased rapidly in the site of the ON vulnerable to secondary degeneration. Increased plasma membrane expression of GluR1 in astrocytes, resulting in increased calcium flux into and across the linked astrocytic network, is thought to lead to the early spread of oxidative stress in neurons (Fitzgerald et al. 2009a).

10.2.3.4 Mitochondrial Change

The alteration in the mitochondrial ultrastructure was observed following PONT injury (Cummins et al. 2013). In addition, the activity of enzymes of the citric acid cycle was dynamically altered during secondary degeneration in the ON (Cummins et al. 2013).

10.2.3.5 C-jun

C-jun was shown to be involved in secondary degeneration after PONT. Western blotting showed that the expression of phospho-c-jun (p-c-jun) increased in the inferior retinas (Li et al. 2013). Immunohistochemical (IHC) staining showed that the number of p-c-jun or c-jun positive cells increased significantly from 3 days after PONT in secondary degeneration (Fitzgerald et al. 2010a; Vander and Levkovitch-Verbin 2012).

10.2.3.6 Water Channel Change

Aquaporin 4 (AQP4) is the main water channel of the mammalian nervous system and is indicated to colocalize with the inward potassium (K) channel Kir 4.1, as a water–potassium transport complex. AQP4 immunointensity in glial fibrillary acidic protein (GFAP)-positive astrocytes increased following PONT (Wells et al. 2012). Although no change in Kir 4.1 immunoreactivity or K ions was observed, it is possible that the change in the flux of ions in response to AQP4-mediated astrocytic swelling/hypertrophy results in the spread of altered ionic balances through the astrocytic syncytium and increased secondary degeneration.

10.2.3.7 Glial Cells

Three days after PONT, the numbers of microglia/macrophages increased significantly in the ventral ON (Fitzgerald et al. 2010a). Astrocytes became hypertrophic after PONT and the immunoreactivity of oxidative stress markers (MnSOD and advanced glycation end product carboxymethyl lysine) increased in the astrocytes of the ventral ON after PONT (Fitzgerald et al. 2010a; Wells et al. 2012). Calcium overload is known to contribute to increased oxidative stress. As mentioned above, the ion channel AMPA receptor subunit GluR1 immunointensity and water channel AQP4 immunointensity in astrocytes was significantly increased in the ventral ON 3 and 24 h following PONT, respectively. Change in AQP4 immunoreactivity may indicate its involvement in altered water balance and flux of ions following PONT. These results indicated astrocytes in the spread of calcium and oxidative stress during secondary degeneration (Fitzgerald et al. 2010a; Wells et al. 2012). Swelling in myelinated axons, but not in unmyelinated axons was detected after PONT in

PVG hooded rats. It implied that the dysfunction of oligodendroglia contributed to axon swelling and damaged axonal transport (Payne et al. 2011). It is known that oligodendrocytes are vulnerable to glutamate excitotoxicity, which may be released from the neural tissue after primary injury. But the occurrence of excitotoxicity after PONT has not been investigated yet and the number of oligodendrocytes remains stable after PONT. Therefore, to understand the role of oligodendrocytes in secondary degeneration, study about excitotoxity should be conducted.

10.2.4   Secondary Degeneration Beyond Ganglion Cell Layer After PONT

What should be noticed is that other retinal layers were also affected after PONT in addition to ganglion cell layer. We investigated the activity of other retinal layers after PONT by electroretinogram (ERG), which is a functional recording of the retinal electrical signals in response to a flash of light received at the retina. The multifocal electroretinogram (mfERG) is an advanced ERG technique to examine the retinal response of multiple loci within a short period of time, which is similar to multiple measurements of localized ERGs. The mfERG measurement provides topographical responses of the retina (Sutter and Tran 1992), which can illustrate the retinal changes caused by glaucomatous damage (Chan and Brown 1999, 2000; Chu et al. 2006, 2007). Previous studies showed that the P1 component of mfERG originated from the outer retinas and maybe ON-bipolar cells in porcine and rhesus monkey (Hood et al. 2002; Ng et al. 2008). Our results showed that the P1 component originated from the outer retina because its amplitude remained unchanged after inhibiting the inner retinal activity by tetrodotoxin (TTX) + N-methyl-D-aspartic acid (NMDA). However, its amplitude reduced across the whole retina after PONT in our study indicated that outer retina was also affected. The damage after PONT is assumed to be limited to the RGC layer, but not other layers. The overall reduction of the P1 component indicates that the secondary degeneration could adversely influence the retinal layers beyond RGC level (Chu et al. 2013). We also noticed, after PONT, the decrease of the mfERG photopic negative response (PhNR) component, which is an electroretinal response related to the inner retina as well as RGC. The PhNR reduction was originally assumed to be found at the site with ON transection, but the general diminish of PhNR of the whole retina illustrates that the secondary degeneration also damages the inner retina not related to the part of ON without transection. This kind of transsynaptic degeneration has been found in glaucoma patients and glaucoma models. For example, the decrease of cone opsin messenger RNA (mRNA) occurred in glaucomatous patients and monkey models (Pelzel et al. 2006), swelling and patchy loss of cone photoreceptors in the macular region (Nork et al. 2000; Calkins 2012). In DBA mice, the best-characterized chronic glaucoma model, detection with retinal histology showed the attenuation of scotopic a- and b-wave amplitude concurrent with thinning of the inner plexiform layer and of the outer retina layer, respectively (Bayer et al. 2001). In C57 mice, the numbers of GABAergic types of amacrine cells decreased by about the same percentages to RGCs (Moon et al. 2005). These data indicated the possible degeneration of photoreceptors and amacrine cells in glaucoma. The mechanisms of this transsynaptic degeneration were not clear, but the degeneration of other layers than the ganglion cell layer should be checked when assessing the effects of a neuroprotective agent.

10.3   L. barbarum

10.3.1   A Brief Introduction of L. barbarum

In Chinese medicine, *L. barbarum* has been used as an "upper class herb" for many years, and it is believed to be good for the vision sight, nourishing the "kidney" and protecting the "liver" (Junlin and Aicheng 2002). It can be eaten directly, and also can be used for making tea, tonic soup, and wine in daily life. *L. barbarum* is made up of many components, for example, polysaccharides (Chang and So 2008; Ho et al. 2009), taurine (Song et al. 2011), betaine (Xie et al. 2001; Lee et al. 2004), zeaxanthin (Kim et al. 2002), beta-carotene (Inbaraj et al. 2008), beta-sitosterol (Xie et al. 2001), flavonoids (Lee et al. 2004), vitamins (Zhang et al. 2011), amino acids (Wang et al. 2012), fatty acids (Honglin et al. 2009), and trace minerals (Yang et al. 2012b).

The biological effects of *L. barbarum* have been reported in different animal models. These studies showed that it had potential benefits against overweight (Amagase and Nance 2011), liver injury (Cui et al. 2011; Xiao et al. 2012), cardiovascular diseases (Xin et al. 2011), inflammatory diseases (Tang et al. 2011), and cancer (Tang et al. 2011; Zhu and Zhang 2012). It could also modulate immunity (Vidal et al. 2012) and sex behavior (Lau et al. 2012). In addition, *L. barbarum* has neuroprotective roles (Yu et al. 2005, 2006; Ho et al. 2007; Chang and So 2008; Ho et al. 2010; Li et al. 2011; Mi et al. 2012b; Chu et al. 2013; Li et al. 2013; He et al. 2014).

10.3.2   The Neuroprotective Effects of L. barbarum

L. barbarum is also known as wolfberry or Gouqizi. The neuroprotective effects of *L. barbarum* have been demonstrated both in vitro and in vivo. *L. barbarum* polysaccharides (LBP) could protect primary cultured hippocampal neurons of neonatal rats from the damage induced by oxygen–glucose deprivation and reperfusion (Rui et al. 2012). In another study, it elucidated the neuroprotective effects of wolfberry against homocysteine-induced neuronal damage (Ho et al. 2010). Glutamate excitotoxicity takes part in many neurodegenerative diseases. The experiments from Chang's group showed that *L. barbarum* could inhibit glutamate neurotoxicity in primary cultures of cortical neurons (Ho et al. 2009). The fruit of *L. barbarum* also elicited an effective protection for neurons against ß-amyloid peptides-induced apoptosis by reducing the activity of both caspase-3 and -2 (Ho et al. 2007). In vivo, research indicated that *L. barbarum* extracts protected the brain from blood-brain barrier disruption and cerebral edema in experimental stroke (Yang et al. 2012a). *L. barbarum* also showed neuroprotective effects on eyes (Chan et al. 2007; Li et al. 2011; Mi et al. 2012b; Chu et al. 2013; Li et al. 2013; He et al. 2014).

10.3.3   The Neuroprotective Effects of L. barbarum on Eyes

L. barbarum has also been studied extensively in eye diseases including diabetic retinopathy, ischemia model, and glaucoma. *L. barbarum* extract enhanced cell viability of a spontaneously arising retinal pigment epithelia line (ARPE-19, a model of diabetic retinopathy) exposed to high glucose injury. This cytoprotective effect was achieved by the reduction of high-glucose-induced apoptosis and downregulation of caspase-3 protein expression (Song et al. 2012). The results from cell culture study have shown that LBP could increase the survival of RGCs from neonatal SD rats (Yang et al. 2011). In animal studies, *L. barbarum* could improve the function of retinas in rats with diabetic retinopathy evaluated by using electroretinogram (Hu et al. 2012). In retinal ischemia/reperfusion model, LBP attenuated neuronal damage, blood-retinal barrier disruption, and oxidative stress (Li et al. 2011). *L. barbarum* could rescue photoreceptors of rd1 mice (Miranda et al. 2010). The human study showed that *L. barbarum* could reduce hypopigmentation and soft drusen accumulation in the macula of elderly subjects (Bucheli et al. 2011b). LBP could decrease the expression of endothelin-1 and modulate the expression of its receptors in a rat OH model (Mi et al. 2012a). It could increase the expression of crystallins and modulate microglia and then protected RGCs in rat hypertension models (Chan et al. 2007; Chiu et al. 2009, 2010).

The protective mechanisms of *L. barbarum* involve antioxidation (Li 2007; Cheng and Kong 2011; Li et al. 2011; Shan et al. 2011; Xiao et al. 2012), anti-excitotoxicity (Ho et al. 2009), anti-inflammation (Wu et al. 2011; Xiao et al. 2012), and anti-apoptosis (Ho et al. 2010; Li et al. 2011; Song et al. 2012).

10.3.4   The Effects of L. barbarum on Secondary
Degeneration of RGCs After PONT

In our previous study using the PONT model of SD rats, it was shown that LBP could reduce secondary degeneration but not primary degeneration of the cell bodies of the RGCs after PONT (Li et al. 2013). This protective effect might be related with the antioxidant effect of LBP and the inhibition of the JNK pathway (Li et al. 2013). In addition to the primary and secondary degeneration of RGCs, the mfERG study showed that outer retinas (photoreceptors and bipolar cells) could also be

Fig. 10.2 The summary of the results from our group about the neuroprotection of LBP on secondary degeneration of retinas after PONT. After PONT, secondary degeneration of RGCs in the inferior retinas is caused by apoptosis attributed to the activation of the JNK pathway, and oxidative stress occurs. The retinal function is damaged by the degeneration of RGCs and secondary functional injury of outer retinas. The administration of LBP can reverse these changes both in structure and function

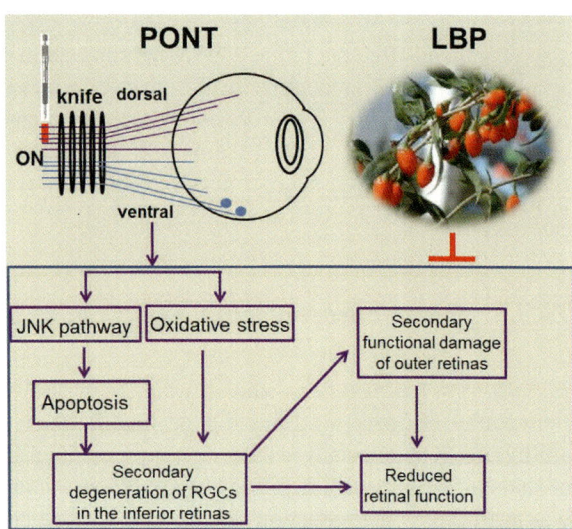

affected subsequently after PONT, and *L. barbarum* could reverse the functional deterioration caused by primary and secondary degeneration of the outer and inner retina (Chu et al. 2013). Therefore, *L. barbarum* could delay secondary degeneration caused by PONT including RGCs and outer retinas. The results from our group are summarized in Fig. 10.2.

10.3.5   The Possible Mechanisms Involved in the Neuroprotection of L. barbarum in Secondary Degeneration

10.3.5.1   Antioxidant Property

MnSOD is an antioxidant enzyme by converting toxic superoxide into hydrogen peroxide and diatomic oxygen. In our study, LBP could increase the production of MnSOD in the retinas after PONT (Li et al. 2013). The study from He et al. showed that LBP could also activate the transcription factor nuclear factor erythroid 2-related factor (Nrf2)/Heme oxygenase-1(HO-1) pathway to exert the antioxidant role in the retina after ischemia/reperfusion damage (He et al. 2014). Both RGCs and amacrine cells lost after ischemia/reperfusion and the degeneration of amacrine cells also occurred in glaucoma (Moon et al. 2005). Therefore, LBP may protect the amacrince cells from secondary degeneration via the antioxidant property.

10.3.5.2   Inhibition of Apoptosis

Apoptosis is involved in the secondary degeneration of RGCs after PONT (Levko-vitch-Verbin et al. 2010; Li et al. 2013). JNKs can bind and phosphorylate c-jun. Three kinds of JNKs, including JNK1, JNK2, and JNK3, are all involved in apopto-sis (Yang et al. 1997; Tournier et al. 2000). JNK3 is found mainly in the neural tis-sues. LBP could reduce the phosphorylation of JNK1 in rat cortical neurons, which were induced toxicity by homocysteine and glutamate (Ho et al. 2009, 2010). Our study showed that LBP could decrease the phosphorylation of JNK3 and c-jun after PONT (Li et al. 2013). Therefore, LBP may inhibit apoptosis via the inhibition of the JNK pathway.

10.3.5.3   Immunomodulatory Effects

LBP could increase the proliferation of lymphocytes in vitro (Bucheli et al. 2011a). In the studies from our group, LBP could modulate the activation of microglia and delay the degeneration of RGCs in chronic rat OH model (Chan et al. 2007).

10.3.5.4   Preserving Synapses

Our study has shown the transsynaptic cell degeneration in retinas after PONT by multifocal ERG testing (Chu et al. 2013). In our group, LBP restored the decreased spine density and the reduced expression of postsynaptic density protein 95 (PSD-95) in the hippocampus (Zhang et al. 2012). PSD-95 exists in the postsynaptic com-ponent of synapses and is important for the maintaining of synapses. Therefore, it is possible that LBP can preserve the structure and function of synapses after PONT, which needs further investigation in the future.

10.3.5.5   Modulating Autophagy

Autophagy is involved in secondary degeneration of axons in the ON (Cummins et al. 2013). LBP could inhibit autophagy after transection of sciatic nerve in rats (Fan et al. 2010). Although the effects of LBP on autophagy have not been investi-gated after PONT, it is a possible mechanism of the neuroprotection of LBP.

10.3.5.6   Modulating Signal Processing

The LBP was found to restore and to enhance the mfERG responses of both outer and inner retina (Chu et al. 2013). This phenomenon was not observed in the eyes without PONT under the supplement of LBP. The protective effect of LBP is be-lieved to be activated in the presence of PONT or other insults. Without any delay of the mfERG components, the recovery of the mfERG may be caused by changing the electrical characteristics of the retinal cells, altering the blood flow and improving the synaptic transmission. Most of the RGCs in the superior retinas degenerated and they are not likely to involve in the restoring of the PhNR after LBP supplement. The influence to the electro-retinal activity may be contributed by other cells in the inner retina, for example, amacrine cells (Kielczewski et al. 2005). Therefore, it seems that after PONT the LBP activate the amacrine cells, which provide an effect of compensation reflex to rescue the signal disturbance from inner retina (or RGCs) by enhancing the outer retinal responses.

Conclusions

In neurodegenerative diseases and other diseases of CNS, secondary degeneration of neurons and glial cells occurred. Protection of neurons and glial cells died from secondary degeneration is a promising direction for therapy of these diseases. PONT model is a useful tool to study the mechanisms of secondary degeneration and to screen the neuroprotective drugs for secondary degeneration since it can separate primary degeneration from secondary degeneration in location. Until now, the mechanisms of secondary degeneration after PONT could be apoptosis, necrosis, and autophagy. Oxidative stress, calcium overload, mitochondria dysfunction, JNK pathway, change of water channel, microglia and macrophages, astrocytes, and oligodendrocytes were involved in the secondary degeneration after PONT. LBP could delay secondary degeneration of RGCs and outer retinas and improve the function of retinas by inhibiting oxidative stress and activation of JNK pathway after PONT. Other possible neuroprotective mechanisms of LBP for secondary degeneration include immunomodulatory effects, preserving synapses, modulating autophagy, and modulating signal processing.

References

Almasieh M, Wilson AM, Morquette B, Cueva Vargas JL, Di Polo A. The molecular basis of retinal ganglion cell death in glaucoma. Prog Retin Eye Res. 2012;31:152–81.

Amagase H, Nance DM. *Lycium barbarum* increases caloric expenditure and decreases waist circumference in healthy overweight men and women: pilot study. J Am Coll Nutr. 2011;30: 304–9.

Bayer AU, Neuhardt T, May AC, Martus P, Maag KP, Brodie S, Lutjen-Drecoll E, Podos SM, Mittag T. Retinal morphology and ERG response in the DBA/2NNia mouse model of angle-closure glaucoma. Invest Ophthalmol Vis Sci. 2001;42:1258–65.

Begemann M, Grube S, Papiol S, Malzahn D, Krampe H, Ribbe K, Friedrichs H, Radyushkin KA, El-Kordi A, Benseler F, Hannke K, Sperling S, Schwerdtfeger D, Thanhauser I, Gerchen MF, Ghorbani M, Gutwinski S, Hilmes C, Leppert R, Ronnenberg A, Sowislo J, Stawicki S, Stodtke M, Szuszies C, Reim K, Riggert J, Eckstein F, Falkai P, Bickeboller H, Nave KA, Brose N, Ehrenreich H. Modification of cognitive performance in schizophrenia by complexin 2 gene polymorphisms. Arch Gen Psychiatry. 2010;67:879–88.

Bucheli P, Gao Q, Redgwell R, Vidal K, Wang J, Zhang W. Biomolecular and clinical aspects of Chinese wolfberry. In: Benzie IFF, Wachtel-Galor S, editors. Herbal medicine: biomolecular and clinical aspects. Boca Raton: Taylor & Francis; 2011a.

Bucheli P, Vidal K, Shen L, Gu Z, Zhang C, Miller LE, Wang J. Goji berry effects on macular characteristics and plasma antioxidant levels. Optom Vis Sci. 2011b;88:257–62.

Calkins DJ. Critical pathogenic events underlying progression of neurodegeneration in glaucoma. Prog Retinal Eye Res. 2012;31:702–19.

Chan HL, Brown B. Multifocal ERG changes in glaucoma. Ophthalmic Physiol Opt. 1999;19: 306–16.

Chan HL, Brown B. Pilot study of the multifocal electroretinogram in ocular-hypertension. Br J Ophthalmol. 2000;84:1147–53.

Chan HC, Chang RCC, Koon-Ching Ip A, Chiu K, Yuen WH, Zee SY, So KF. Neuroprotective effects of *Lycium barbarum* Lynn on protecting retinal ganglion cells in an ocular hypertension model of glaucoma. Exp Neurol. 2007;203:269–73.

Chang RCC, So KF. Use of anti-aging herbal medicine, *Lycium barbarum*, against aging-associated diseases. What do we know so far? Cell Mol Neurobiol. 2008;28:643–52.

Cheng D, Kong H. The effect of *Lycium barbarum* polysaccharide on alcohol-induced oxidative stress in rats. Molecules. 2011;16:2542–50.

Chiu K, Chan HC, Yeung SC, Yuen WH, Zee SY, Chang RCC, So KF. Modulation of microglia by wolfberry on the survival of retinal ganglion cells in a rat ocular hypertension model. J Ocul Biol Dis Infor. 2009;2:47–56.

Chiu K, Zhou Y, Yeung SC, Lok CK, Chan OO, Chang RCC, So KF, Chiu JF. Up-regulation of crystallins is involved in the neuroprotective effect of wolfberry on survival of retinal ganglion cells in rat ocular hypertension model. J Cell Biochem. 2010;110:311–20.

Chu PHW, Chan HHL, Brown B. Glaucoma detection is facilitated by luminance modulation of the global flash multifocal electroretinogram (mfERG). Invest Ophthalmol Vis Sci. 2006;47:929–37.

Chu PHW, Chan HHL, Brown B. Luminance-modulated adaptation of global flash mfERG: fellow eye losses in asymmetric glaucoma. Invest Ophthalmol Vis Sci. 2007;48:2626–33.

Chu PH, Li HY, Chin MP, So KF, Chan HH. Effect of *Lycium barbarum* (wolfberry) *polysaccharides* on preserving retinal function after partial optic nerve transection. PLoS One. 2013;8:e81339.

Cui B, Liu S, Lin X, Wang J, Li S, Wang Q, Li S. Effects of *Lycium barbarum* aqueous and ethanol extracts on high-fat-diet induced oxidative stress in rat liver tissue. Molecules. 2011;16: 9116–28.

Cummins N, Bartlett CA, Archer M, Bartlett E, Hemmi JM, Harvey AR, Dunlop SA, Fitzgerald M. Changes to mitochondrial ultrastructure in optic nerve vulnerable to secondary degeneration in vivo are limited by irradiation at 670 nm. BMC Neurosci. 2013;14:98.

Fan H, Deng C, Fu J, Ding L, Yin G, Ma Y. [Effects of *Lycium barbarum* polysaccharide on formation of traumatic neuroma and pain after transection of sciatic nerve in rats]. Zhongguo Xiu Fu Chong Jian Wai Ke Za Zhi. 2010;24:1298–301.

Fitzgerald M, Bartlett CA, Evill L, Rodger J, Harvey AR, Dunlop SA. Secondary degeneration of the optic nerve following partial transection: the benefits of lomerizine. Exp Neurol. 2009a;216:219–30.

Fitzgerald M, Payne SC, Bartlett CA, Evill L, Harvey AR, Dunlop SA. Secondary retinal ganglion cell death and the neuroprotective effects of the calcium channel blocker lomerizine. Invest Ophthalmol Vis Sci. 2009b;50:5456–62.

Fitzgerald M, Bartlett CA, Harvey AR, Dunlop SA. Early events of secondary degeneration after partial optic nerve transection: an immunohistochemical study. J Neurotrauma. 2010a;27: 439–52.

Fitzgerald M, Bartlett CA, Payne SC, Hart NS, Rodger J, Harvey AR, Dunlop SA. Near infrared light reduces oxidative stress and preserves function in CNS tissue vulnerable to secondary degeneration following partial transection of the optic nerve. J Neurotrauma. 2010b;27:2107–19.

Fitzgerald M, Hodgetts S, Van Den Heuvel C, Natoli R, Hart NS, Valter K, Harvey AR, Vink R, Provis J, Dunlop SA. Red/near-infrared irradiation therapy for treatment of central nervous system injuries and disorders. Rev Neurosci. 2013;24:205–26.

Guimaraes JS, Freire MA, Lima RR, Souza-Rodrigues RD, Costa AM, dos Santos CD, Picanco-Diniz CW, Gomes-Leal W. Mechanisms of secondary degeneration in the central nervous system during acute neural disorders and white matter damage. Rev Neurol. 2009;48:304–10.

Hausmann ON. Post-traumatic inflammation following spinal cord injury. Spinal Cord. 2003;41:369–78.

He M, Pan H, Chang RCC, So KF, Brecha NC, Pu M. Activation of the Nrf2/HO-1 antioxidant pathway contributes to the protective effects of Lycium barbarum polysaccharides in the rodent retina after ischemia-reperfusion-induced damage. PLoS One. 2014;9:e84800.

Ho YS, Yu MS, Lai CS, So KF, Yuen WH, Chang RCC. Characterizing the neuroprotective effects of alkaline extract of Lycium barbarum on beta-amyloid peptide neurotoxicity. Brain Res. 2007;1158:123–34.

Ho YS, Yu MS, Yik SY, So KF, Yuen WH, Chang RCC. Polysaccharides from wolfberry antagonizes glutamate excitotoxicity in rat cortical neurons. Cell Mol Neurobiol. 2009;29:1233–44.

Ho YS, Yu MS, Yang XF, So KF, Yuen WH, Chang RCC. Neuroprotective effects of polysaccharides from wolfberry, the fruits of Lycium barbarum, against homocysteine-induced toxicity in rat cortical neurons. J Alzheimers Dis. 2010;19:813–27.

Honglin S, Nengjun X, Qian G, Chaomin N. Analysis of fatty acids components of medlar by GC/MS. J Chin Mass Spectrom Soc. 2009;30:99–104.

Hood DC, Frishman LJ, Saszik S, Viswanathan S. Retinal origins of the primate multifocal ERG: implications for the human response. Invest Ophthalmol Vis Sci. 2002;43:1673–85.

Hu CK, Lee YJ, Colitz CM, Chang CJ, Lin CT. The protective effects of Lycium barbarum and chrysanthemum morifolum on diabetic retinopathies in rats. Vet Ophthalmol. 2012;15:65–71.

Inbaraj BS, Lu H, Hung CF, Wu WB, Lin CL, Chen BH. Determination of carotenoids and their esters in fruits of Lycium barbarum linnaeus by HPLC-DAD-APCI-MS. J Pharm Biomed Anal. 2008;47:812–18.

Junlin L, Aicheng W. Gou qi. Beijing: Bejing Science and Technology Press; 2002.

Kielczewski JL, Pease ME, Quigley HA. The effect of experimental glaucoma and optic nerve transection on amacrine cells in the rat retina. Invest Ophthalmol Vis Sci. 2005;46:3188–96.

Kim HP, Lee EJ, Kim YC, Kim J, Kim HK, Park JH, Kim SY, Kim YC. Zeaxanthin dipalmitate from Lycium chinense fruit reduces experimentally induced hepatic fibrosis in rats. Biol Pharm Bull. 2002;25:390–92.

Lau BW, Lee JC, Li Y, Fung SM, Sang YH, Shen J, Chang RCC, So KF. Polysaccharides from wolfberry prevents corticosterone-induced inhibition of sexual behavior and increases neurogenesis. PLoS One. 2012;7:e33374.

Lee CJ, Lee JH, Seok JH, Hur GM, Park Js J, Bae S, Lim JH, Park YC. Effects of betaine, coumarin and flavonoids on mucin release from cultured hamster tracheal surface epithelial cells. Phytother Res. 2004;18:301–05.

Levkovitch-Verbin H, Quigley HA, Kerrigan-Baumrind LA, D'Anna SA, Kerrigan D, Pease ME. Optic nerve transection in monkeys may result in secondary degeneration of retinal ganglion cells. Invest Ophthalmol Vis Sci. 2001;42:975–82.

Levkovitch-Verbin H, Quigley HA, Martin KR, Zack DJ, Pease ME, Valenta DF. A model to study differences between primary and secondary degeneration of retinal ganglion cells in rats by partial optic nerve transection. Invest Ophthalmol Vis Sci. 2003;44:3388–93.

Levkovitch-Verbin H, Dardik R, Vander S, Melamed S. Mechanism of retinal ganglion cells death in secondary degeneration of the optic nerve. Exp Eye Res. 2010;91:127–34.

Levkovitch-Verbin H, Spierer O, Vander S, Dardik R. Similarities and differences between primary and secondary degeneration of the optic nerve and the effect of minocycline. Graefes Arch Clin Exp Ophthalmol. 2011;249:849–57.

Lewen A, Matz P, Chan PH. Free radical pathways in CNS injury. J Neurotrauma. 2000;17:871–90.

Li XM. Protective effect of Lycium barbarum polysaccharides on streptozotocin-induced oxidative stress in rats. Int J Biol Macromol. 2007;40:461–65.

Li SY, Yang D, Yeung CM, Yu WY, Chang RCC, So KF, Wong D, Lo AC. *Lycium barbarum* polysaccharides reduce neuronal damage, blood-retinal barrier disruption and oxidative stress in retinal ischemia/reperfusion injury. PLoS One. 2011;6:e16380.

Li H, Liang Y, Chiu K, Yuan Q, Lin B, Chang RCC, So KF. *Lycium barbarum* (wolfberry) reduces secondary degeneration and oxidative stress, and inhibits JNK pathway in retina after partial optic nerve transection. PLoS One. 2013;8:e68881.

Mi X, Chiu K, Van G, Leung JWC, Lo ACY, Chung SK, Chang RCC, So K. Effect of Lycium barbarum Polysaccharides on the expression of endothelin-1 and its receptors in an ocular hypertension model of rat glaucoma. Neural Regen Res. 2012a;7:645–51.

Mi XS, Feng Q, Lo AC, Chang RCC, Lin B, Chung SK, So KF. Protection of retinal ganglion cells and retinal vasculature by *Lycium barbarum* polysaccharides in a mouse model of acute ocular hypertension. PLoS One. 2012b;7:e45469.

Miranda M, Arnal E, Ahuja S, Alvarez-Nolting R, Lopez-Pedrajas R, Ekstrom P, Bosch-Morell F, van Veen T, Romero FJ. Antioxidants rescue photoreceptors in rd1 mice: relationship with thiol metabolism. Free Radic Biol Med. 2010;48:216–22.

Moon JI, Kim IB, Gwon JS, Park MH, Kang TH, Lim EJ, Choi KR, Chun MH. Changes in retinal neuronal populations in the DBA/2J mouse. Cell Tissue Res. 2005;320:51–9.

Ng YF, Chan HH, Chu PH, Siu AW, To CH, Beale BA, Gilger BC, Wong F. Pharmacologically defined components of the normal porcine multifocal ERG. Documenta ophthalmologica. Adv Ophthalmol. 2008;116:165–76.

Nickells RW. From ocular hypertension to ganglion cell death: a theoretical sequence of events leading to glaucoma. Can J Ophthalmol. 2007;42:278–87.

Nork TM, Ver Hoeve JN, Poulsen GL, Nickells RW, Davis MD, Weber AJ, Vaegan, Sarks SH, Lemley HL, Millecchia LL. Swelling and loss of photoreceptors in chronic human and experimental glaucomas. Arch Ophthalmol. 2000;118:235–45.

Oyinbo CA. Secondary injury mechanisms in traumatic spinal cord injury: a nugget of this multiply cascade. Acta Neurobiol Exp (Wars). 2011;71:281–99.

Payne SC, Bartlett CA, Harvey AR, Dunlop SA, Fitzgerald M. Chronic swelling and abnormal myelination during secondary degeneration after partial injury to a central nervous system tract. J Neurotrauma. 2011;28:1077–88.

Payne SC, Bartlett CA, Harvey AR, Dunlop SA, Fitzgerald M. Myelin sheath decompaction, axon swelling, and functional loss during chronic secondary degeneration in rat optic nerve. Invest Ophthalmol Vis Sci. 2012;53:6093–101.

Payne SC, Bartlett CA, Savigni DL, Harvey AR, Dunlop SA., Fitzgerald M. Early proliferation does not prevent the loss of oligodendrocyte progenitor cells during the chronic phase of secondary degeneration in a CNS white matter tract. PLoS One. 2013;8:e65710.

Pelzel HR, Schlamp CL, Poulsen GL, Ver Hoeve JA, Nork TM, Nickells RW. Decrease of cone opsin mRNA in experimental ocular hypertension. Mol Vis. 2006;12:1272–82.

Rui C, Yuxiang L, Yinju H, Qingluan Z, Yang W, Qipeng Z, Hao W, Lin M, Juan L, Chengjun Z, Yuanxu J, Yanrong W, Xiuying D, Wannian Z, Tao S, Jianqiang Y. Protective effects of *Lycium barbarum* polysaccharide on neonatal rat primary cultured hippocampal neurons injured by oxygen-glucose deprivation and reperfusion. J Mol Histol. 2012;43:535–42.

Savigni DL, O'Hare Doig RL, Szymanski CR, Bartlett CA, Lozic I, Smith NM, Fitzgerald M. Three Ca channel inhibitors in combination limit chronic secondary degeneration following neurotrauma. Neuropharmacology. 2013;75C:380–90.

Selt M, Bartlett CA, Harvey AR, Dunlop SA, Fitzgerald M. Limited restoration of visual function after partial optic nerve injury; a time course study using the calcium channel blocker lomerizine. Brain Res Bull. 2010;81:467–71.

Shan X, Zhou J, Ma T, Chai Q. *Lycium barbarum* polysaccharides reduce exercise-induced oxidative stress. Int J Mol Sci. 2011;12:1081–88.

Song MK, Salam NK, Roufogalis BD, Huang TH. *Lycium barbarum* (Goji Berry) extracts and its taurine component inhibit PPAR-gamma-dependent gene transcription in human retinal pigment epithelial cells: possible implications for diabetic retinopathy treatment. Biochem Pharmacol. 2011;82:1209–18.

Song MK, Roufogalis BD, Huang TH. Reversal of the caspase-dependent apoptotic cytotoxicity pathway by taurine from *Lycium barbarum* (Goji Berry) in human retinal pigment epithelial cells: potential benefit in diabetic retinopathy. Evid Based Complement Alternat Med. 2012;2012:323784.

Stewart SS, Appel SH. Trophic factors in neurologic disease. Annu Rev Med. 1988;39:193–201.

Stoica BA, Faden AI. Cell death mechanisms and modulation in traumatic brain injury. Neurotherapeutics. 2010;7:3–12.

Sutter EE, Tran D. The field topography of ERG components in man – I. The photopic luminance response. Vison Research. 1992; 32, 433–446.

Szymanski CR, Chiha W, Morellini N, Cummins N, Bartlett CA, O'Hare Doig RL, Savigni DL, Payne SC, Harvey AR, Dunlop SA, Fitzgerald M. Paranode abnormalities and oxidative stress in optic nerve vulnerable to secondary degeneration: modulation by 670 nm light treatment. PLoS One. 2013 8:e66448.

Tang WM, Chan E, Kwok CY, Lee YK, Wu JH, Wan CW, Chan RY, Yu PH, Chan SW. A review of the anticancer and immunomodulatory effects of *Lycium barbarum* fruit. Inflammopharmacology. 2011

Tezel G. Oxidative stress in glaucomatous neurodegeneration: mechanisms and consequences. Prog Retin Eye Res. 2006;25:490–513.

Tezel G. TNF-alpha signaling in glaucomatous neurodegeneration. Prog Brain Res. 2008;173:409–21.

Tournier C, Hess P, Yang DD, Xu J, Turner TK, Nimnual A, Bar-Sagi D, Jones SN, Flavell RA, Davis RJ. Requirement of JNK for stress-induced activation of the cytochrome c-mediated death pathway. Science. 2000;288:870–74.

Vander S, Levkovitch-Verbin H. Regulation of cell death and survival pathways in secondary degeneration of the optic nerve—a long-term study. Curr Eye Res. 2012;37:740–48.

Vidal K, Bucheli P, Gao QT, Moulin J, Shen LS, Wang JK, Blum S, Benyacoub J. Immunomodulatory effects of dietary supplementation with a milk-based wolfberry formulation in healthy elderly: a randomized, double-blind, placebo-controlled trial. Rejuvenation Res. 2012;15:89–97.

Wang Y, Zhao H, Han F. Determination of amino acids and trace elements in aqueous extracts from *Lycium barbarum* L. Food Ind. 2012;33:113–15.

Wells J, Kilburn MR, Shaw JA, Bartlett CA, Harvey AR, Dunlop SA, Fitzgerald M. Early in vivo changes in calcium ions, oxidative stress markers, and ion channel immunoreactivity following partial injury to the optic nerve. J Neurosci Res. 2012;90:606–18.

Wu PS, Wu SJ, Tsai YH, Lin YH, Chao JC. Hot water extracted *Lycium barbarum* and *Rehmannia glutinosa* inhibit liver inflammation and fibrosis in rats. Am J Chin Med. 2011;39:1173–91.

Xiao J, Liong EC, Ching YP, Chang RCC, So KF, Fung ML, Tipoe GL. *Lycium barbarum* polysaccharides protect mice liver from carbon tetrachloride-induced oxidative stress and necroinflammation. J Ethnopharmacol. 2012;139:462–70.

Xie C, Xu LZ, Li XM, Li KM, Zhao BH, Yang SL. Studies on chemical constituents in fruit of *Lycium barbarum* L. Zhongguo Zhong Yao Za Zhi. 2001;26:323–24.

Xin Y, Zhang S, Gu L, Liu S, Gao H, You Z, Zhou G, Wen L, Yu J, Xuan Y. Electrocardiographic and biochemical evidence for the cardioprotective effect of antioxidants in acute doxorubicin-induced cardiotoxicity in the beagle dogs. Biol Pharm Bull. 2011;34:1523–26.

Yang DD, Kuan CY, Whitmarsh AJ, Rincon M, Zheng TS, Davis RJ, Rakic P, Flavell RA. Absence of excitotoxicity-induced apoptosis in the hippocampus of mice lacking the Jnk3 gene. Nature. 1997;389:865–70.

Yang M, Gao N, Zhao Y, Liu LX, Lu XJ. Protective effect of *Lycium barbarum* polysaccharide on retinal ganglion cells in vitro. Int J Ophthalmol. 2011;4:377–79.

Yang D, Li SY, Yeung CM, Chang RCC, So KF, Wong D, Lo AC. *Lycium barbarum* extracts protect the brain from blood-brain barrier disruption and cerebral edema in experimental stroke. PLoS One. 2012a;7:e33596.

Yang RM, Suo YR, Wang HL. [Determination and analysis of trace elements in *Lycium barbarum* L. from different regions of Qinghai province]. Guang Pu Xue Yu Guang Pu Fen Xi. 2012b;32:525–28.

Yoles E, Schwartz M. Degeneration of spared axons following partial white matter lesion: implications for optic nerve neuropathies. Exp Neurol. 1998;153:1–7.

Yu MS, Leung SK, Lai SW, Che CM, Zee SY, So KF, Yuen WH, Chang, RCC. Neuroprotective effects of anti-aging oriental medicine *Lycium barbarum* against beta-amyloid peptide neurotoxicity. Exp Gerontol. 2005;40:716–27.

Yu MS, Ho YS, So KF, Yuen WH, Chang RCC. Cytoprotective effects of *Lycium barbarum* against reducing stress on endoplasmic reticulum. Int J Mol Med. 2006;17:1157–61.

Zhang Z, Liu X, Zhang X, Liu J, Hao Y, Yang X, Wang Y. Comparative evaluation of the antioxidant effects of the natural vitamin C analog 2-O-beta-D-glucopyranosyl-L-ascorbic acid isolated from goji berry fruit. Arch Pharm Res. 2011;34:801–10.

Zhang E, Yau SY, Lau BW, Ma H, Lee TM, Chang RCC, So KF. Synaptic plasticity, but not hippocampal neurogenesis, mediated the counteractive effect of wolfberry on depression in rats(1). Cell Transplant. 2012;21:2635–49.

Zhu CP, Zhang SH. *Lycium barbarum* polysaccharide inhibits the proliferation of HeLa cells by inducing apoptosis. J Sci Food Agric. 2012.

Chapter 11
Role of *Lycium Barbarum* Extracts in Retinal Diseases

María Benlloch, María Muriach, Gloria Castellano, Francisco Javier Sancho-Pelluz, Emilio González-García, Miguel Flores-Bellver and Francisco J. Romero

Abstract Wolfberry (*Lycium barbarum*) extracts have been used in the treatment of some retinal degenerative diseases, mostly associated with other antioxidants. The ones this chapter focuses on are retinitis pigmentosa and age-related macular degeneration. Hereditary retinal dystrophies are a broad (and growing) group of hereditary disorders affecting the retina. Retinitis pigmentosa is perhaps the best known of them and is sometimes (inaccurately) used as a synonym for some of the other conditions in this category. Although these two entities show completely different etiologies: The first one is a group of hereditary diseases in which photoreceptor death occurs due to different mutations, and in the second the loss of photoreceptors is associated to the aging process, in both oxidative damage has been claimed as a pathophysiological mechanism. The present chapter reviews the antioxidant chemical features of wolfberry and wolfberry extracts, the oxidative mechanisms involved in the pathophysiology of photoreceptor cell death, and the existing data on the use of wolfberry for retinal degenerations.

Keywords Wolfberry · Goji berries · Age-related macular degeneration (AMD) · Diabetic retinopathy (DR) · Retinitis pigmentosa (RP) · Glutathione (GSH) · Oxidative stress · Antioxidant · *Lycium barbarum* polysaccharides (LPBs)

Dedicated to Prof. Theo van Veen on the ocassion of his 70th birthday.

F. J. Romero (✉) · M. Benlloch · G. Castellano · F. J. Sancho-Pelluz · E. González-García · M. Flores-Bellver
Universidad Católica de Valencia 'San Vicente Mártir', Valencia, Spain
e-mail: fj.romero@ucv.es

F. J. Romero
Department of Physiology, School of Medicine, Universidad Católica de Valencia 'San Vicente Mártir', 46001 Valencia, Spain
e-mail: fj.romero@ucv.es

M. Muriach
Unidad predepartamental Medicina, Universitat Jaume I, Castellón de la Plana, Spain

© Springer Science+Business Media Dordrecht 2015
R. C-C. Chang, K-F. So (eds.), *Lycium Barbarum and Human Health*,
DOI 10.1007/978-94-017-9658-3_11

153

11.1   Wolberry Extracts: Properties and Healthy Benefits

11.1.1   *Lycium Barbarum Fruit: Chemical Compounds of Wolberry*

Lycium barbarum (LB) is a member of the *Solanaceae* family that has bright red fruits. Those red-colored fruits, also called *Fructus lycii*, Gouqizi, Goji berry, or wolfberry, have been used as a traditional Chinese herbal medicine for thousands of years (Xin et al. 2011). LB is a species closely related to *Lycium chinese*. Both are medicinal plants native to China but are also widely found in Korea, Japan, and other Asian countries (Zhong et al. 2013).

Some of the compounds that could be found in this species are:

- *Essential oils:* hexadecanoic acid (47.5%), linoleic acid (9.1%), β-elemene (5.4%), myristic acid (4.2%), and ethyl hexadecanoate (4.0%) (Altintas et al. 2006).
- *Vitamins and amino acids*: It contains one of the largest quantities of vitamin C compared to any other fruit in the world. It also contains a number of other vitamins like A, B1, B2, B6, E as well as 18 different amino acids including essential ones.
- *Polysaccharides*: rhamnose, xylose, mannose, arabinose, and galactose.
- *Minerals*: iron, copper, zinc, calcium, potassium, phosphor, and selenium.
- *Carotenoids*: Zeaxanthin dipalmitate, β-cryptoxanthin monopalmitate and its two isomers, zeaxanthin monopalmitate and its two isomers, all-trans-β-carotene, and all-trans-zeaxanthin (Inbaraja et al. 2008).
- *Flavonoids*: Polyphenols are important active compounds in LB, since highly efficient extraction techniques, such as ultrasonic-assisted extraction, microwave-assisted extraction, and a combination of microwave and ultrasonication in the extraction of compounds have been reported to extract biologically active substances from plant materials (Dong et al. 2009). The contents of total flavonoids (21.25 mg/g) of cultivated LB leaves is much higher than those in the wild type LB leaves (17.86 mg/g), so cultivated LB leaves are a suitable source for medicine vegetables and functional tea.

The predominant flavonoids in LB were identified as rutin and quercetin-diglycoside, quercetin, kaempferol, and kaempferol-3-O-rutinoside. Moreover, it contains phenolic compounds: chlorogenic acid, caffeoylquinic acid, caffeic acid, and p-coumaric acid (Wang et al. 2010). Their varieties and amount vary because of differences in growing environments, maturity, and growth conditions (Erlund 2004).

11.2   LB Fruits: Wolfberry Healthy Compounds

LB fruits have a large variety of beneficial effects and play an important role in preventing many different diseases. In fact, in 1983 The Ministry of The Public Health of China approved LB fruits to be marketed as a botanical medicine (Shan et al. 2011).

Many functional components in LB, including flavonoids, carotenoids, and polysaccharides, have been reported to be closely associated with their health-enhancing effect.

Before summarizing the beneficial effects of LB in general, it is important to indicate that there are some studies that described wolfberries as new allergenic sources. Goji berries are described as responsible for a large number of allergic sensitizations, mainly in patients residing in the Mediterranean area, ranging from oral allergy symptoms to anaphylactic reactions (Carnés et al. 2013). The role of lipid transfer proteins from Goji berry has been demonstrated after analyzing the allergenic profile of sensitized individuals and cross-reactivity studies with other allergen extracts (Carnés et al. 2013).

11.2.1   *Beneficial Effects of Polyphenols from Goji Berries*

Polyphenols are among the most widespread class of secondary metabolites in nature. Most polyphenols arise from a common origin: the amino acids phenylalanine or tyrosine. These amino acids are deaminated to cinnamic acids, which enter the phenylpropanoid pathway. A key step in this biosynthetic route is the introduction of one or more hydroxyl groups into the phenyl ring. They appear to function as protectors against various biotic and abiotic stresses, e.g., oxidative stress (Hammerbacher et al. 2011; He et al. 2008; Rosemann et al. 1991).

As mentioned above, some of the flavonoids contained in LB like quercetin, kaempferol, and rutin may be relevant to the following actions.

There is much interest in the biological effects of flavonoids because they exhibit antioxidant, anti-inflammatory, antitumor, antimutagenic, antiviral and antiallergical, antibacterial (bactericidal, bacteriostatic), algicidal, antifungal, insecticidal, estrogenic, and keratolytic activities (Pietta 2000; Lacikova et al. 2009; Khadem and Marles 2010; Chong et al. 2010; Yang et al. 2001). Moreover, flavonoids may play an important role in atherosclerosis prevention and therapy (see *other important effects*). The preventive effects of these secondary plant metabolites in terms of cardiovascular, neurodegenerative diseases, and cancer are deduced from epidemiologic as well as *in vitro* and *in vivo* data and result in the corresponding nutritional recommendations. Furthermore, polyphenols were found to modulate the activity of a wide range of enzymes (involved in radical generation) and cell receptors. In this way, in addition to having antioxidant properties, polyphenols have several other specific biological actions in preventing and/or treating diseases. Many reports have described the importance of structure–bioactivity relationship of flavonoids and phenolic compounds (Castellano et al. 2012a, b; Castellano et al. 2013).

11.2.1.1 Antioxidant Activity

Flavonoids can act as antioxidants by a number of potential pathways. The most important is likely to be related to their free radical scavenging activity, in which the polyphenol can break the free radical chain reaction.

The radical-scavenging antioxidants inhibit the free-radical-mediated oxidation of lipids, proteins, and DNA, which is involved in disease. Several methods have been designed to measure the activity of compounds, such as the lipid peroxidation inhibition capacity (LPIC) assay, 1,1-diphenyl-2-picrylhydrazyl (DPPH) scavenging (Simic et al. 2007), cyclic voltammetry (Kadoma and Fujisawa 2008), and the induction period method (Omata et al. 2008). The assessment of oxidative damage to proteins can involve measurement of specific amino acid adducts with malondialdehyde (MDA), or derived from the attack of reactive oxygen or nitrogen species (e.g., hydroxytyrosine or nitrotyrosine).

A number of studies have been carried out on the structure—antioxidant activity relationships of the flavonoids. The importance of antioxidant activity in flavonoids' structures follow a hierarchical order (Castellano et al. 2013): either cinnamic or benzoic ester group at 3-position on ring C (as epigallocatechingallate); *ortho*-dihydroxy (catechol) structure on the B ring; the presence of two hydroxyl groups at positions 3 and 5; the presence of a 2,3-double bond in conjugation with a 4-keto function on the B ring; a 2,3-double bond and 4,5-double bond; the absence of alkoxyl and glycoxyl ester groups on ring B; and the absence of alkoxyl and glycoxyl ester groups on ring A. Flavonoids contained in LB are characterized by the presence of a 2,3-double bond in conjugation with a 4-keto function on the B ring, providing electron delocalization from the B ring and molecular planarity, stabilizing the radical generated. Hydroxyl group at position 5 provides hydrogen bonding to the keto group. Rutin presents an *ortho*-dihydroxy (catechol) structure on the B ring, the most important factor in the antioxidant character because of the importance that this structural fragment has to scavenge free radicals, as it is widely described in the technical literature. These structural features are illustrated in Fig. 11.1.

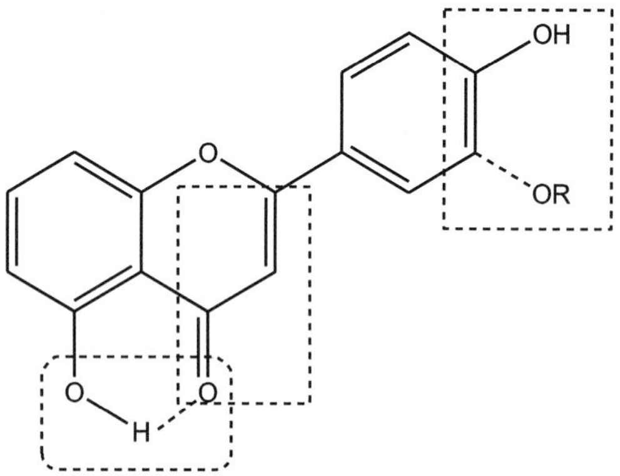

Fig. 11.1 Structural features of *L. barbarum* flavonoids

Moreover, rutin and kaempferol-3-O-rutinoside present absence of alkoxyl and glycoxyl ester groups on ring A. These facts explain the high antioxidant potency of them together with the fact that antioxidants act synergistically together with other antioxidants *in vivo* (Natella et al. 1999).

Another pathway of apparent antioxidant action of the flavonoids, particularly in oxidation systems using transition metal ions such as copper or iron, is chelation of the metal ions. Chelations of catalytic metal ions may prevent their involvement in Fenton-type reactions, which can generate highly reactive hydroxyl radicals.

$$H_2O_2 + Cu^+ \rightarrow OH + OH^- + Cu^{2+}$$

$$Cu^{2+} + O_2 \rightarrow Cu^+ + O_2$$

The ability of polyphenols to react with metal ions may also render them prooxidants. The possible prooxidant effects of flavonoids may be important *in vivo* if free transition metal ions are involved in oxidation processes. In the healthy human body, metal ions appear largely sequestered in forms unable to catalyze free radical reactions. However, injury to tissues may release iron or copper, and catalytic metal ions have been measured in atherosclerotic lesions. In these cases the potential of flavonoids to act as prooxidants may increase (Croft 1998).

Phenolic acids may also be good antioxidants. They act as such in a number of ways. Phenolic hydroxyl groups are good hydrogen donors: hydrogen-donating antioxidants can react with reactive oxygen and reactive nitrogen species in a termination reaction, which breaks the cycle of new radical's generation.

Following interaction with the initial reactive species, a radical form of the antioxidant is produced, having a much greater chemical stability than the initial radical. The interaction of the hydroxyl groups of phenolics with the π-electrons of the benzene ring gives the molecules special properties, most notably the ability to generate free radicals where the radical is stabilized by delocalization. The formation of these relatively long-lived radicals is able to modify radical-mediated oxidation processes (Pereira et al. 2009).

The naturally occurring phenolic acids share the frame structure of hydroxycinnamic acid or hydroxybenzoic acid and are present in many foods and plants. Research interests also arise for food quality since they are associated with color, sensory qualities, and nutritional properties. Recent interest has focused on their antioxidant properties and potential health. The antioxidant activity of phenolic acids is related to the number and position of hydroxyl groups in the molecule. The antioxidant efficiency of mono-phenols is strongly enhanced by the introduction of a second methoxyl or hydroxyl group at the *o*- or *p*-position with respect to the hydroxyl above (Castellano et al. 2012b; Szwajgier et al. 2005). According to previous reports, the introduction of a second hydroxyl group in the *o*-position (caffeic acid, chlorogenic acid, caffeoylquinic acid) or *p*-position (protocatechuic acid) enhances the antioxidant activity, making these phenolic acids more efficient than their respective monofenols (*p*-coumaric, syringic acid). This is consistent with the electron withdrawing potential of the single carboxyl functional group on the phenol ring affecting the *o*- and *p*-positions. Caffeic acid (hydroxycinnamic acid) is the most antioxidant compound. This is in agreement with the literature, in which hydroxycinnamic acids were found to be more effective than their hydroxybenzoic acid counterparts, possibly because of the aryloxy-radical stabilizing effect of the –CH=CH–COOH linked to the phenyl ring by resonance.

Moreover, we predict that the LB extract has an important antioxidant character, since it contains in addition to flavonoids, a great number of phenolic acid derivatives (benzoics and cinnamics), especially chlorogenic acid and *p*-coumaric acid, caffeic acid, and caffeoylquinic acid. However, in stress situations, the composition of phenols increases because these are formed to protect the plant from reactive oxygen species (ROS), anthropogenic pressures, and interspecific competition.

UV Absorption of the Main Flavonoids Flavonoids as rutin, the main LB flavonoid, are synhetisized in higher plants to protect them from the harmful effects of UV-B radiation and diseases (Dong et al. 2009). The increased Mean Increasing Ratio of Absorbance (MIRA) value of rutin under UV radiation indicated that rutin has the capacity of anti-UV and that cultivated LB leaves might be good sources for anti-radiation food or anti-UV cosmetics and protection to retinal diseases.

11.2.1.2   Antitumor Activity

Cancer is a multistep disease incorporating environmental, chemical, physical, metabolic, and genetic factors. It was found that in addition to their primary antioxidant activity, this group of compounds displays a wide variety of biological functions which are mainly related to modulation of carcinogenesis.

Natural phenolics can affect basic cell functions that are related to cancer development by many different mechanisms (Dai and Mumper 2010). They may limit the formation of the initiated cells by stimulating DNA repair (Webster et al. 1996).

Secondly, phenolics may inhibit the formation and growth of tumors by induction of cell cycle arrest and apoptosis (Dai and Mumper 2010). Polyphenols have been found to affect cancer cell growth by inducing apoptosis in many cell lines such as the hepatoma (HepG2), the colon (SW620, HT-29, CaCo-2, and HCT-116), the prostate (DU-145 and LNCaP), the lung (A549), the breast (MCF-7), the melanoma (SK-MEL-28 and SK-MEL-1), the neuroblastoma (SH-SY5Y), and the HL-60 leukemia cells.

Recently, many *in vitro* studies have been published on the modulation of oncogenes, tumor-suppressor genes, cell cycle, apoptosis, angiogenesis, and related signal transduction pathways by polyphenols (Yang et al. 2001).

11.2.1.3   Other Important Activities

There are some anti-inflammatory effects for LB flavonoids described, such as, for example, inhibition of the expression of intercellular adhesion molecule-1

(ICAM-1) and vascular cell adhesion molecule (VCAM-1) induced by tumor necrosis factor-α (TNF-α) in human umbilical vein endothelial cells (HUVECs) (Wu et al. 2012), and anti-atherosclerosis.

These anti-inflammation and antioxidant effects are studied together by researchers in the context of age-related diseases, including neurodegeneration, diabetes (Chang and So 2008), and increase of immune responses in the elderly (especially vaccine response) (Vidal et al. 2012).

11.2.2   Beneficial Effects of Carotenoids from Goji Berries

The presence of two free carotenoids and seven carotenoid esters has been reported in wolfberries. Of the various carotenoids, the presence of zeaxanthin and its esters dominate the amount of carotenoids in wolfberries.

A high content of carotenoids provide high provitamin A value and antioxidant activity (Lin et al. 2011b). Zeaxanthin dipalmitate is the predominant carotenoid in Goji berries (Peng et al. 2005), followed by L-cryptoxanthin monopalmitate, zeaxanthin monopalmitate, L-carotene, and zeaxanthin (Inbaraja et al. 2008).

Zeaxanthin and lutein (isomeric dihydroxycarotenoids) are the major constituents of the retinal macular region (Bone and Landrum 1992). There are many studies showing the effect of Zeaxanthin in age-related macular degeneration (AMD) (see Sect. 11.4).

Zeaxanthin and other carotenoids are also potent antioxidants, which contribute to the health effects of Goji berries against the oxidative stress-mediated diseases (Zhong et al. 2013). Kim et al. studied the antihepatotoxic activity of zeaxanthin dipalmitate, showing a significant hepatoprotective activity against carbon tetrachloride toxicity and exerting a potent hepatoprotective activity by inhibiting Ito cell proliferation, collagen synthesis, and by inhibiting certain biochemical functions of Kupffer cells (Kim et al. 1997).

There are some studies about the effect of carotenoids such as zeaxanthin in cardiovascular diseases. For example, patients with coronary artery disease showed lower plasma levels of lutein, zeaxanthin, β-cryptoxanthin, α-carotene, β-carotene, and lycopene compared to healthy subjects. Moreover, the reduced levels of lutein, zeaxanthin, and β-cryptoxanthin were associated with smoking, high body mass index, and low high-density lipoprotein cholesterol (HDL-C) (Lidebjer et al. 2007). The antihypertensive effect of carotenoids is supported by a follow-up study in which the concentrations of the sum of four serum carotenoids (α-carotene, β-carotene, lutein/zeaxanthin, and cryptoxanthin) were inversely correlated with incident hypertension after 20 years (Hozawa et al. 2009).

Similar to other botanical (plant and fungal-derived) polysaccharides, LB polysaccharides (LBP) mainly occur as water-soluble glycoconjugates, for example, conjugates of glycan with peptides or proteins (Zhong et al. 2013).

In the last few years, the polysaccharides isolated from the aqueous extracts of LB fruits are described as one of the most valuable functional constituents and active compounds responsible for various health effects (Shan 2011).

Many studies on pharmacology and phytochemistry have demonstrated that LBP had various bioactivities such as antioxidant, immunomodulation, antitumor, antidiabetic, etc. (Jin et al 2013).

11.2.2.1   Antioxidant Activities

Many of the biological activities of LBP are directly or indirectly attributed to their antioxidant potential as many chronic diseases are oxidative stress-mediated (Zhong et al. 2013).

Antioxidants are substances that help reduce the severity of oxidative stress either by forming a less active radical or by quenching the reaction. It has been reported that LBP may possess the capacity to donate hydrogen to superoxide anion because of the weak dissociation energy of O–H bond (Jin et al. 2011).

Shan et al. analyzed the effects of LBP on exercise-induced oxidative stress in rats, showing that serum levels of MDA in an LBP-treated group were significantly decreased compared with that in the normal control group. Superoxide dismutase (SOD) and glutathione peroxidase (GPx) activities of rats (hind-limb skeletal muscle) in LBP-treated groups were significantly increased compared with that in the normal control (Shan et al. 2011).

In a similar study, Zhao et al. (2013) investigated the effects of LBP on arterial compliance during exhaustive exercise (swimming exercise) in rats. The rats administered LBPs showed longer swimming time until exhaustion than the control group rats. Exercise-induced MDA elevation was repressed by LBPs supplementation.

The LBPs significantly upregulated the expression of endothelial nitric oxide (NO) synthase (eNOS) and improved the endothelium-dependent vasodilatation of the aorta ring, showing that LBPs administration significantly inhibited the oxidative stress and improved the arterial compliance.

Along that same line, many authors described the effects of LBP in the decrease of MDA and beneficial increase of several antioxidant enzymes activities suggesting that the antioxidant activity of LBP is mainly attributed to the improvement of antioxidant enzymatic activities more than the donation of hydrogen to superoxide anion (Jin et al. 2011).

Some examples are the study of Amagase and Nance (2008) that investigated the antioxidant effects of LBP on 50 Chinese healthy adults aged 55–72 years in a 30-day randomized, double-blind, placebo-controlled clinical study. The results showed that LBP treatment significantly increased the serum levels of SOD by 8.4% and GPx by 9.9% and serum MDA decreased by 8.7%, indicating that LBP could support health in humans by stimulating endogenous factors and protecting membranes from oxygen radical-mediated damage.

Recently, other antioxidant effects were studied (Chen et al. 2008). In this study, authors investigated the therapeutic effects of LBPs on learning and memory and neurogenesis in scopolamine (SCO)-treated rats. SCO administration led to damage of dendritic development of new neurons. LBP prevented these SCO-induced reductions in cell proliferation and neuroblast differentiation. LBPs decreased the SCO-induced oxidative stress in hippocampus and reversed the ratio Bax/Bcl-2 that was increased after SCO treatment. Those results suggest that suppression of oxidative stress and apoptosis may be involved in the above effects of LBPs that may be a promising candidate to restore memory functions and neurogenesis.

11.2.2.2   Antidiabetic Activities

Diabetes is associated with significant oxidative stress, and increasing evidences suggested that oxidative stress caused by hyperglycemia plays an important role in the pathogenesis of diabetes mellitus (Jin et al. 2012).

In 2004, Luo et al. studied the effects of purified LBP fractions in alloxan-induced diabetic or hyperlipidemic rabbits showing that those LBP fractions reduced blood glucose levels, total cholesterol (TC), and triglyceride (TG) concentrations at the same time that HDL levels markedly increased after 10 days of treatment in tested rabbits (Luo et al. 2004).

Along this line, Cui et al. (2011) and Wu et al. (2010) evaluated the effects of LBP on blood lipid metabolism, blood glucose, and oxidative stress of mice fed with high-fat diet. They found that LBP significantly decreased the levels of low-density lipoprotein (LDL), TC, TG, and blood glucose, and increased the activities of SOD, GPx, and catalase compared with high-fat diet groups.

11.2.2.3   Antitumor Effects

Some evidences have demonstrated that the anticancer activities of different chemotherapeutic agents are involved in the induction of apoptosis, which is regarded as the preferred way to manage cancer (Hsu et al. 2004). Zhang et al. (2005) investigated the effects of LBP on the proliferation rate, cell cycle distribution, and apoptosis in human hepatoma QGY7703 cell line. LBP treatment caused the inhibition of cell growth with cycle arrest in S phase and apoptosis induction. Mao et al. (2011) revealed that LBP treatment inhibited the growth of two different colon cancer cell lines (SW480 and Caco-2 cell lines) in a dose-dependent manner and cells were arrested at the G0/G1 phase.

The antitumor activity of LBPs seems to come from the induction of cell-cycle arrest and apoptosis, and inhibition of some signalling pathways, which play a protective effect against carcinogenesis by eliminating abnormal excess of tumor cells (Jin et al 2013).

11.3   Oxidative Mechanisms Involved in the Pathophysiology of Photoreceptor Cell Death in Retinal Diseases

The retina, located at the back of the eyecup, is a multilayer tissue with the main purpose of performing phototransduction: to transform photonic signals into electric energy that the brain is able to understand (Kolb 2003). Retinal cells can be altered in many ways and oxidative stress is known to play an important role in the development of several retinal diseases such as AMD, diabetic retinopathy (DR), and retinitis pigmentosa (RP) (Berson et al. 1993; Ganea and Harding 2006; Bazan 2006).

Even though the retina is part of the central nervous system and is also affected by oxidative stress (Halliwel 1992), it seems that photoreceptors are especially sensitive to oxidation due to its anatomical location and its function (Tanito et al. 2002). For that reason, survival of photoreceptors and other retinal cells depends on appropriate antioxidant mechanisms. Glutathione (GSH), together with different GSH-related enzymes, belongs to this antioxidant defense (Ganea and Harding 2006).

Retinitis Pigmentosa RP is a group of heterogeneous inherited retinal degeneration that results in photoreceptor cell death (Hartong et al. 2006). Usually, rod photoreceptor cells are affected by a mutation and degeneration in certain time, whereas cone photoreceptors die secondarily even though they are not affected by the mutation (Sancho-Pelluz et al. 2008). RP produces nyctalopia or night blindness and reduction of peripheral vision (tunnel vision) which might develop to loss of central vision. Even though the primary cell death is, in most of the cases, well understood, secondary cell death of photoreceptors is still under study. Different mechanisms are discussed as causes of mutation-independent cell death of cones. Changes in oxygen consumption seem to be important in the rapid development of the disease. Loss of rods can be translated into a reduction of oxygen consumption and therefore a hyperoxia in the photoreceptor layer and an increase in oxidative stress in the remaining cells, both cones and surviving rods (Shen et al. 2005; Wellard et al. 2005). As seen earlier, ROS are normal products of mitochondrial metabolism. Protective mechanisms that include antioxidants and repair mechanisms have been confirmed to be active in the human retina (Puertas et al. 1993). An amplification of oxidative stress or a failure of the defense mechanisms leads to changes in metabolism which ultimately, if not resolved, may result in cell degeneration (Marcum et al. 2005). Oxidative stress has been implicated in the pathogenesis of RP. It was demonstrated that antioxidant therapies were able to slow down cell death rate in animal models of cone and rod degeneration (Komeima et al. 2006; Sanz et al. 2007). It is clear that redox status is imbalanced in a number of retinal maladies. In fact, in the majority of photoreceptor disorders, rod photoreceptor cell death is due to the genetic mutation, but cones, which are not affected by the mutation, must die in a mechanistically different way (Punzo et al. 2009). Lately, it has been demonstrated that mitochondria oxidative stress is an important source of ROS in different animal models of inherited photoreceptor degeneration (Vlachantoni et al. 2011).

The *rd1* mouse, probably the best-known animal model for RP to date, holds a mutation in the gene that encodes for phosphodiesterase 6 (PDE6), producing a nonfunctional PDE6, essential for phototransduction (Bowes et al. 1990). Rod photoreceptor degeneration begins at postnatal day (P) 10, and is almost complete at P21; only cone cells remain at the photoreceptor layer. However, those cones will eventually die due to secondary nonmutation-related degeneration. An involvement of oxidative stress was found in this animal model for both mutation-induced rod cell death (Sanz et al. 2007) and secondary cone cell death (Komeima et al. 2006). Moreover, antioxidants were found to be protective *in vitro* studies of photoreceptor degeneration (Chucair et al. 2007).

Augmented oxidative stress may originate from either excessive energy demand, impairment of oxidative phosphorylation, or reduced antioxidant defenses. It has been hypothesized that Ca^{2+} concentration is elevated in the cytosol of *rd1* rod cells. This might be caused because mutated PDE6 produces excessive accumulation of cyclic guanosine monophosphate (cGMP), and therefore the cGMP-activated cyclic nucleotide gated (CNG) ion channels are continuously open, leading to an Na^+ and a Ca^{2+} accumulation (Paquet-Durand et al. 2009). Elevation of intracellular Ca^{2+} produces the activation of mitochondrial calpain-10, which opens mitochondrial permeability transition pores (MPTP) (Arrington et al. 2006), which could explain translocation of apoptosis inducing factor (AIF) from mitochondria to the nucleus (Artus et al. 2006). Given that AIF has considerable redox activity, calpain-induced depletion of mitochondrial AIF would result in mitochondrial dysfunction and increased oxidative stress (Yamashima 2004).

It has been observed that the *rd1* mouse also presents a decrease of defensive mechanisms against oxidative stress such as the action of GSH-S-transferase and GPx (Ahuja et al. 2008), where such downregulation is likely to enhance the deleterious effects of ROS.

11.3.1 Diabetic Retinopathy

It has been repeatedly suggested that oxidative stress is involved in the pathogenesis of late diabetes complications (Baynes and Thorpe 1996), though it is not definitely demonstrated if this is the cause or the consequence of these complications. It is clear that the elevated glucose levels present in diabetes and the existence of oxidative stress are inseparable (Packer et al 2001). The retina is extremely rich in polyunsaturated lipid membranes and this feature makes it especially sensitive to oxygen- and/or nitrogen-activated species and lipid peroxidation.

Hyperglycemia reduces antioxidant levels and concomitantly increases the production of free radicals; and in fact, in animal models of diabetes mellitus, reduced levels of retinal antioxidants, such as SOD, GSH reductase, and GPx, as well as nonenzymatic antioxidants such as vitamins C and E, and beta-carotene, have been observed. Levels of GSH, a key scavenger of ROS, are also reduced in the retina of diabetic rats (Muriach et al. 2006). Moreover, superoxide levels are elevated in the retina of diabetic rats and in retinal cells incubated in high-glucose media (Kowluru and Abbas 2003; Du et al. 2003; Cui et al. 2006), and hydrogen peroxide content is increased in the retina of diabetic rats (Ellis et al. 2000). Membrane lipid peroxidation and oxidative damage to DNA (indicated by 8-hydroxy-2-deoxyguanosine, 8-OHdG), the consequences of ROS-induced injury, are elevated in the retina in diabetes (Kowluru and Abbas 2003; Ellis et al 2000; Muriach et al. 2006; Miranda et al. 2004, 2006, 2007). These effects contribute to tissue damage in diabetes mellitus, leading to alterations in the redox potential of the cell with subsequent activation of redox-sensitive genes (Bonnefont-Rousselot 2002). Moreover, oxidative stress is linked to early apoptosis in DR both at the microvasculature and neuronal cells of the retina, but oxidative stress also appears to be highly interrelated with other biochemical imbalances that lead to structural and functional changes (Madsen-Bouterse and Kowluru 2008). Thus, for example, it is well known that chronic overproduction of ROS in the retina results in aberrant mitochondrial functions in diabetes (Kowluru 2005).

Among the proposed pathogenic mechanisms, the polyol pathway model has received the most scrutiny. Aldose reductase (AR) is the first enzyme in the polyol pathway, converting excess glucose to sorbitol, which is then metabolized to fructose by sorbitol dehydrogenase. According to several studies, AR is responsible for the early events in the pathogenesis of DR, leading to a cascade of retinal lesions including blood-retinal barrier (BRB) breakdown, loss of pericytes, neuroretinal apoptosis, glial reactivation, and neovascularization (Caldwell et al. 2005). Increased AR activity has been shown to contribute to increased oxidative stress by promoting nonenzymatic glycation and the activation of protein kinase C (PKC) (Stitt and Curtis 2005). It has been demonstrated that AR inhibition counteracts diabetes-induced oxidative and nitrosative stress and prevents vascular endothelial growth factor (VEGF) overexpression, basement membrane thickening, pericyte loss, and microaneurysms in retinal capillaries (Obrosova et al. 2003). In long-term diabetes-induced neuroretinal stress, increased expression of VEGF and apoptosis and proliferation of blood vessels have been shown to be less prominent than in AR-deficient animals (Obrosova et al. 2005).

A recent clinical study has substantiated the concept of hyperglycemic memory in the pathogenesis of DR. The Diabetes Control and Complications Trial-Epidemiology of Diabetes Interventions and Complications Research, has revealed that the reduction in the risk of progressive retinopathy resulting from intensive therapy in patients with type 1 diabetes persisted for at least several years after the Diabetes Control and Complications Trial (DCCT), despite increasing hyperglycemia. The process of formation and accumulation of advanced glycation end products (AGEs) and their mode of action are most compatible with the theoretical hyperglycemic memory (Yamagishi et al 2008). AGEs are formed by nonenzymatic reactions between reducing sugars and free amino groups of proteins or lipids. AGEs have been detected within retinal vasculature and neurosensory tissue of diabetic eyes. Multiple consequences of AGEs accumulation in the retina have been demonstrated, including upregulation of VEGF, upregulation of NF-κB, and increased leukocyte adhesion in retinal microvascular endothelial cells (Lu et al. 1998; Moore et al. 2003).

In a 5-year study in diabetic dogs, administration of aminoguanidine (an inhibitor of AGEs formation) prevented retinopathy (Kern and Engerman 2001). AGEs exert cell-mediated effects via AGEs receptor (RAGE), a multiligand signal-transduction receptor of the immunoglobulin superfamily (Schmidt et al. 1992). Consequences of ligand-RAGE interaction include increased expression of VCAM-1, vascular hyperpermeability, enhanced thrombogenicity, induction of oxidant stress, and abnormal expression of eNOS (Schmidt et al. 1995; Wautier et al. 1996). Recently, it has been shown that after RAGE activation nicotinamide adenine dinucleotide phosphate (NADPH) oxidase is activated by phospholipase C-mediated activation of Ca^{2+}-dependent PKC and that this may lead to an increase in ROS that could be associated with the initial stages of macular edema and DR (Warboys et al. 2005). Studies in models of retinopathy show that increases in oxidative stress and signs of vascular inflammation are correlated with increases in arginase activity and arginase 1 expression; decreasing arginase expression or inhibiting its activity blocks these effects; and that the induction of arginase during retinopathy is blocked by inhibiting NADPH oxidase activity (Caldwell et al. 2010). Finally, it has been also demonstrated that AGEs can induce glial reaction and neuronal degeneration in retinal explants (Lecleire-Collet et al. 2005).

Alterations associated with oxidative stress offer many potential therapeutic targets making this an area of great interest to the development of safe and effective treatments for DR. Animal models of DR have shown beneficial effects of antioxidants on the development of retinopathy, but clinical trials (though very limited in numbers) have provided somewhat ambiguous results. Although antioxidants are being used for other chronic diseases, controlled clinical trials are warranted to investigate potential beneficial effects of antioxidants in the development of retinopathy in diabetic patients.

Age-Related Macular Degeneration   AMD is the major cause of blindness in the elderly with over 1.7 million people having reduced vision due to this disorder in the USA (Friedman et al. 2004). The disease affects the macula at the center of the eye and as a consequence results in loss of central vision which significantly impacts the patient's ability to read, watch television, or drive. The major pathological changes associated with AMD are observed in the functionally and anatomically related tissues, including photoreceptors, retinal pigment epithelium (RPE), Bruch's membrane, and choriocapillaries (Bhutto and Lutty 2012). AMD is broadly divided into two forms: dry and wet, that account for about 85 and 15 % of cases, respectively. Wet AMD, the most severe form of AMD, is generally associated with subretinal (i.e., between the retina and choroid) neovascularization and considerable amelioration can be achieved with the use of antiangiogenic agents (Andreoli and Miller 2007). Dry macular degeneration is mainly clinically diagnosed based on the presence of drusen, thickening of Bruch's membrane and RPE progressive degeneration. Drusen are deposits of yellowish waste products from photoreceptors cells, which compromise their nutrition inducing ischemia and cell death, which in turn give rise to atrophy and finally results in irreversible visual loss (geographic atrophy). This disorder appears to consist of both a genetic and environmental component with a number of gene polymorphisms being identified which increase susceptibility to environmental risk factors such as smoking, hypertension, diet, or oxidative stress (Ding et al. 2009; Khandhadia and Lotery 2010; Montezuma et al. 2007; Swaroop et al. 2009; Ting et al. 2009). However, unequivocal proof of oxidative stress as a major causative factor in AMD, is difficult due to the complex nature of AMD and its restriction, in the true form of the disease, to humans (Beatty et al. 2001; Winkler et al. 1999; Zarbin 2004). In any case, there is considerable evidence to support oxidative stress as a contributing factor in the onset and progression of AMD as reviewed by Jarrett and Boulton (2012), with mitochondria representing a major source of endogenous ROS in the photoreceptors and underlying RPE (Jarrett et al. 2008; Liang and Godley 2003).

Recent studies show that pro-inflammatory cytokines, TNF-α, interleukin-1β, and interferon-γ, induce ROS in RPE cells via mitochondria and NADPH oxidase (Yang et al. 2007). The cross talk between NADPH oxidases and mitochondria may represent a "vicious cycle" of ROS production with mitochondria being a target for NADPH oxidase-generated ROS and mitochondrial ROS under certain conditions may stimulate NADPH oxidases. Moreover, studies have now begun to provide evidence supporting a role for polymorphic genes associated with oxidative stress at various stages of AMD (Jarrett and Boulton 2012), and recent studies have also implicated important roles for specific microRNAs in AMD. Thus, the miRNA, mir-23 is associated with increased RPE cellular resistance to oxidative stress and was found to be significantly downregulated in macular RPE isolated from AMD patients (Lin et al. 2011a). On the other hand, although light and oxygen are essential for vision they can also lead to photo-induced ROS and subsequent photochemical damage to the retina. The retina contains a variety of chromophores, which being excited at the appropriate wavelength can lead to significant photochemical damage, being the two major photosensitizers the visual pigments in photoreceptor cells and lipofuscin, which accumulates with age in the RPE (Boulton et al. 2001). Finally, other photoreactive molecules in the retina that can generate ROS under certain conditions include melanin, hemoglobin and other iron-containing proteins (e.g., cytochrome C), flavins, flavoproteins, and carotenoids (Boulton et al. 2001).

11.4 Evidences and Future Applications of Wolfberry Extracts in Retinal Diseases

Wolfberries extracts are widely used for the treatment of ocular disorders (Yang and Gao 2011). Most of the reports about the beneficial effects of LB compounds in different retinal diseases are studies about LB polysaccharides and carotenoids but not about polyphenol compounds.

Those research reports are mainly about LBPs and carotenoids effects on ischemia/reperfusion (I/R) injury (including glaucoma as a possible consequence), DR, and AMD. LBPs exert beneficial effects in animal models of ocular diseases. In fact, LBP extracts, are important water-soluble components which have protective effects in promoting survival and prolonging growth of retinal ganglion cells (RGCs) (Yang and Gao 2011) and they have been shown to protect RGCs in an animal model of chronic ocular hypertension (OH) (Chan et al. 2007).

In this part of the chapter, we review current and future applications of Goji berries in those retinal diseases.

11.4.1 Ischemia/Rerfusion Injury

There are many causes of retinal ischemia, including vein and artery occlusions, macular degeneration, diabetes, glaucoma, hypertension, etc. One of the complications after retinal I/R injuries is oxidative stress, which plays a role due to the high content of polyunsatured fatty acids in retina. ROS produced during I/R facilitate lipid peroxidation of membranes, denaturation of proteins, and DNA damage. Oxidative injury is accompanied by retinal swelling, neuronal cell death, and glial cell activation (He et al. 2014).

Many reports show the relationship between LBPs beneficial effects and retinal ischemia and the mechanisms of action proposed are many.

In this line, Li et al. studied the protective effects of LBPs against retinal I/R injury. They focus on three aspects apart from antioxidant activity: anti-apoptosis, preservation of BRB integrity and prevention of retinal swelling (Li et al. 2011). In this study, the authors treated mice with LBPs for 1 week prior to induction of ischemia and showed that LBP effectively protect the retina from neuronal death, apoptosis, glial cell activation, aquaporin water channel upregulation, disruption of BRB, and oxidative stress in retinal I/R induced by surgical occlusion of the internal carotid artery, 1 week after treatment with polysaccharides. Another study demonstrated that one possible mechanism for the protection of the retina by LBPs was immune modulation, which is an indirect effect; neuroprotective effects of LBPs were partly due to modulation of microglia activation (Chiu et al. 2009).

Cells have antioxidant defence systems to counteract oxidative stress. NF-E2-related factor 2/Antioxidant response element (Nrf2/ARE) pathway is one of the antioxidant pathways involved in counteracting increased oxidative stress and maintaining the redox status in many tissues (Ziaei et al. 2013; Kansanen et al. 2013).

One of the enzymes regulated by Nrf2/ARE is heme oxygenase (HO-1), which catalyzes the degradation of heme to biliverdin, CO, and iron. It is demonstrated that pharmacological induction of HO-1 protects the retina from acute glaucoma-induced ischemia-reperfusion injury (Sun et al. 2010).

He et al. studied the relation between the protection of LBPs against retinal damage induced by ischemia-reperfusion injury and activation of the Nrf2/HO-1 pathway. They saw that LBP pretreatment, not only reduced the generation of ROS, but also enhanced the activation of the Nrf2/HO-1 antioxidant pathway in I/R retinas. Furthermore, inhibition of HO-1 activity significantly blocked LBP-induced protective effects on I/R retinas, suggesting that the protective effects of LBP in I/R were mediated at least partly, by the activation of the Nrf2/HO-1 antioxidant pathway (He et al. 2014).

In this line, Chiu et al. suggest that LBPs have a neuroprotective effect on the survival of RGCs mediated via direct upregulation of neuronal survival signal βB2-crystallin. Crystallins are prominent proteins both in normal retina and in retinal diseases. These authors observed that rats fed with LBPs or with PBS did not change the crystallin profiles in the nonstress retinas. But the group, where chronic OH was induced and that was fed with LBPs, presented an increase of crystallin expression of more than tenfold, when compared with control (Chiu et al. 2010).

Mi et al. studied the protection of RGCs and retinal vasculature by LBPs in a chronic OH model (COH) of rat glaucoma. They observed a decrease in the expression of endothelin-1 (a potent vasoconstrictor synthesized in vascular endothelial cells) and its receptors (ET_A and ET_B) under COH condition (Mi et al. 2012a).

Acute OH (AOH) is another well-established animal model of retinal degeneration (Scarsella et al. 2012; Pescosolido et al. 1998). Mi et al. studied the protection of RGCs and retinal vasculature by LBPs in an AOH model suggesting a possible mechanism that consisted in downregulating the RAGE expression (when it is overexpressed in blood vessel, endothelial cells can produce the accumulation of amyloid-ß) and endothelin-1, as well as the related signalling pathways leading to amelioration of vascular damage and neuronal degeneration in AOH (Mi et al. 2012b).

LBPs have shown a beneficial effect on retinal astrocytes and Müller cells in a model of middle cerebral artery occlusion, resulting in the stabilization of the microenvironment that supports the survival of neurons in the retina and brain (Mi et al. 2012b).

11.4.1.1 Diabetic Retinopathy

Hyperglycemia-linked oxidative stress and/or consequent endoplasmic reticulum (ER) stress are part of the pathogenesis of DR. Dietary bioactive components which mitigate oxidative stress may serve as potential chemopreventive agents to prevent or slow disease progression (Tang et al. 2011).

Song et al. investigated the effects of an LB extract on an *in vitro* model of DR as: the retinal ARPE-19 cell line exposed to high glucose (Song et al. 2012). They demonstrated that LB extract is cytoprotective against high-glucose cytotoxicity in ARPE-19 cells, at least in part by regulating apoptosis as a result of caspase-3 modulation. They used taurine in their study at concentrations present in the extracts of LB and they found that the effects of the extract were closely mimicked by taurine. For that reason, they suggest the therapeutic use of taurine and the valuable medicinal herb LB for the prevention of DR.

In this line, Pavan et al. studied the effects of wolfberries and taurine in another human RPE (HRPE) cells. Increased intracellular glucose stimulates the activity of cytosolic adenylyl cyclase, causing an increase of cytosolic cyclic adenosine monophosphate (cAMP) level, leading to a destabilization of epithelial barrier and they found that the protective effect of LB extracts is mediated by the ability to reverse cytosolic adenylyl cyclase stimulation by glucose, thereby avoiding increased cytosolic cAMP levels (Pavan et al. 2014). In this way, they confirm the Song's data (Song et al. 2012) showing that LB extract and its main component, taurine, are able to counteract high glucose-induced RPE damage.

Mitochondria biogenesis, and their metabolic activity, would be altered in diabetes as a consequence of hyperglycemia-induced cellular oxidative stress (Zhang et al. 2012). In fact, prolonged inhibition of AMP-activated protein kinase (AMPK) causes mitochondrial dysfunction and disruption of cellular ROS homeostasis (Kukidome et al. 2006).

Tang et al. studied the effects of dietary wolfberries at early stages of type 2 diabetes in db/db mice (model of the db/db leptin receptor deficient type 2 diabetic mouse) (Tang et al. 2011). They concluded that, taken together, dietary wolfberries and/or its bioactive constituents zeaxanthin and lutein, functioned as modulators of cell survival/death signalling pathways, through targeting pathways in AMPK and Forkhead O transcription factor 3 (FOXO3) signalling, resulting in normalization of cellular ROS and attenuation of ER stress. Activation of FOXO3 protects cells from oxidative stress in diabetes (Li et al. 2009).

Yu et al. studied the effects of dietary wolfberry in db/db mice too. They demonstrated, among other things, that wolfberry elevated zeaxanthin and lutein levels in the liver and retinal tissues; induced activation and nuclear enrichment of retinal AMPK-2; attenuated hypoxia and mitochondrial stress by declining expression of hypoxia-inducible factor-1-, VEGF, and heat shock protein 60; enhanced retinal mitochondrial biogenesis in diabetic retinas, as demonstrated by reversed mitochondrial dispersion in the RPE; increased mitochondrial copy number; elevated citrate synthase activity; and upregulated expression of peroxisome proliferator-activated receptor γ co-activator 1, nuclear respiratory factor 1, and mitochondrial transcription factor A (Yu et al 2013).

Hu et al. describe the effects of LB extracts on rat DR through electroretinography. The electroretinographic amplitudes of the a-waves (that reflect the general physiological *status* of the photoreceptors in the outer retina) and b-waves (that reflect the health of the inner layers of the retina) were significantly decreased in the diabetic animals. However, reductions in the a- and b-wave amplitudes were not observed in the LB-treated group suggesting that LB may have protective effects in DR (Hu et al. 2012).

11.4.1.2   Age-Related Macular Degeneration

Although AMD is a primary cause of vision loss in the elderly, unfortunately, there is no cure for AMD and current treatment options have limited effectiveness and introduce significant patient risk. Therefore, AMD prevention strategies should be identified and implemented. A number of AMD risk factors have been determined including aging, clinical family history, cigarette smoking, white race, and low dietary intake of antioxidants in the form of fruits and vegetables. Given that cigarette smoking and low antioxidant intake are the only modifiable risk factors, increasing antioxidant intake may arguably be the most easily used AMD prevention strategy (Bucheli et al. 2011).

Zeaxanthin intake increases plasma zeaxanthin concentrations, which subsequently increases preretinal pigment concentration, i.e., lutein and zeaxanthin, and ultimately lowers AMD risk (Beatty et al. 2001; Carpentier et al. 2009). More concretely, in Age-Related Eye Disease Study (AREDS), 4519 participants increased dietary lutein/zeaxanthin intake (as determined by a food frequency questionnaire at enrolment) was inversely associated with prevalent neovascular AMD, geographical atrophy, and large or extensive intermediate drusens (Age-Related Eye Disease Study Research Group 2007). In addition, in 2006, the AREDS concluded that lutein-rich and zeaxanthin-rich diets might protect against intermediate AMD in female patients less than 75 years of age (Moeller et al. 2006). More recently, the Blue Mountain Eye study reported that higher dietary lutein and zeaxanthin intake reduced the risk of incidence of early or neovascular AMD over 5 and 10 years (Tan et al. 2008).

Since Goji berry is the richest natural source of the antioxidant carotenoid, zeaxanthin, regular consumption of Goji berry could play a role in the prevention of AMD and help to maintain preretinal pigment density. Although Chinese consumers believe in the benefits of Goji berry on eyesight, very few human studies on the effects of Goji berry supplementation on visual parameters have been reported. Thus, although it has been reported that general endogenous antioxidant markers in serum increased with 30 days of Goji berry juice supplementation (Amagase et al. 2009), a 15-day regimen of Goji berry juice supplementation had no effect on visual acuity in healthy young adults (Amagase and Nance 2008). Another study, however, reported that daily dietary supplementation with Goji berry for 90 days increased plasmatic zeaxanthin and antioxidants levels as well as protected from hypopigmentation and soft drusen accumulation in the macula of elderly subjects (Bucheli et al. 2011). There is a clear relationship between macula hypopigmentation and soft drusen number with AMD risk (Spraul and Grossniklaus 1997).

Our group has studied the effect of Retinacomplex® (a diet supplement enriched with 300 mg LBPs, 50 mg alpha-lipoic acid, 100 mg L-GSH, 10 mg Lutein and 5 mg Zeaxanthin) in AMD patients (they took two Retinacomplex® pills per day, during 24 months). We found that the Retinacomplex® group of patients had an improvement in macular functionality after 2 years of treatment. We observed that presence of diabetes could be an important factor about the AMD evolution, showing better responses of the multifocal electroretinogram in the Retinacomplex® group than the control (González 2013).

In view of these results, it has been proposed that wolfberry may help to halt the progression of AMD. In this sense, two published randomized control trials using *Ginkgo biloba* extracts in AMD showed positive effects on vision. Thus, Lebuisson et al. reported a statistically significant improvement in long distance visual acuity of patients with senile macular degeneration after the treatment with *Ginkgo biloba* extract (Lebuisson et al. 1986). Also Fies et al. showed the therapeutic efficacy of *Ginkgo biloba* extract in patients with senile, dry macular degeneration (Fies and Dienel 2002). However, as discussed by Evans, those trials were small and suffered from short observation periods (Evans 2013), and well-designed and large-scaled clinical trials for the use of Chinese Herbal Medication to prevent or retard progression of AMD are largely lacking (Manheimer et al. 2009).

In conclusion although the role of wolfberry and its extracts seems to be promising for reversing oxidative stress and possibly ameliorating AMD, physicians should be aware that convincing evidence of their dosage and efficacy in AMD, in relationship to safety, tolerability, pharmacokinetics, and pharmacodynamics, is still lacking.

11.4.1.3   Retinitis Pigmentosa

RP is a group of inherited disorders characterized by progressive photoreceptor degeneration leading to night blindness, peripheral vision loss, and subsequently central vision loss. Oxidative stress has been implicated in the pathogenesis of RP, and several studies have demonstrated that antioxidants are able to decrease photoreceptor cell death in different mouse models of RP (Komeima et al. 2006; Miranda et al. 2007). Nowadays, there is almost no study in the literature showing the effects of LB in this disease. In fact, it is noteworthy that only our research group has reported recently that a mixture of antioxidants, including LB, were able to rescue photoreceptors in *rd1* mice (Miranda et al. 2010). *Rd1* mice have a mutation in the same gene that has been found in some human forms of autosomal recessive RP, making the *rd1* mouse retina an ideal model for experimental analysis of human retinal dystrophies (McLaughlin et al. 1995). Thus, we treated *rd1* mice daily from P3 until P11, by oral infusion with a mix that included LBPs (175 mg/kg body wt.). Interestingly, after that treatment GPx activity and GSH levels were increased and cysteine concentrations in *rd1* retinas were decreased. Moreover, significant decreases in the number of transferase uridyl nick end labeling (TUNEL)- and avidin-positive cells were observed (Miranda et al. 2010). Preliminary data of our group with the diet supplement Retinacomplex® (mentioned above) have showed that the patients treated stopped disease progression (in terms of ERG parameters) during the 2 years of Retinacomplex® administration (unpublished data). These results support a possible beneficial role of LB in preventing the progression of RP; however, new studies focused on LB alone are necessary to elucidate this possible beneficial role.

References

Altintas A, Kosar M, Kirimer N, Baser KHC, Dermici B. Composition of the essential oils of *Lycium barbarum* and *L. ruthenicum* fruits. Chem Nat Compd. 2006;42:24–5.

Ahuja S, Ahuja-Jensen P, Johnson LE, Caffé AR, Abrahamson M, Ekström PA, van Veen T. rd1 Mouse retina shows an imbalance in the activity of cysteine protease cathepsins and their endogenous inhibitor cystatin C. Invest Ophthalmol Vis Sci. 2008;49:1089–96.

Amagase H, Nance DM. A randomized, double-blind, placebocontrolled, clinical study of the general effects of a standardized *Lycium barbarum* (Goji) Juice, GoChi. J Altern Complement Med. 2008;14:403–12.

Amagase H, Sun B, Borek C. *Lycium barbarum* (goji) juice improves *in vivo* antioxidant biomarkers in serum of healthy adults. Nutr Res. 2009;29:19–25.

Andreoli CM, Miller JW. Anti-vascular endothelial growth factor therapy for ocular neovascular disease. Curr Opin Ophthalmol. 2007;18:502–8.

Artus C, Maquarre E, Moubarak RS, Delettre C, Jasmin C, Susin SA, Robert-Lézénès J. CD44 ligation induces caspase-independent cell death via a novel calpain/AIF pathway in human erythroleukemia cells. Oncogene. 2006;25:5741–51.

Arrington DD, Van Vleet TR, Schnellmann RG. Calpain 10: a mitochondrial calpain and its role in calcium-induced mitochondrial dysfunction. Am J Physiol Cell Physiol. 2006;291:C1159–71.

Baynes JW, Thorpe SR. The role of oxidative stress in diabetic complications. Curr Opin Endocrinol. 1996;3:277–84.

Bazan NG. Cell survival matters: docosahexaenoic acid signalling, neuroprotection and photoreceptors. Trends Neurosci. 2006;29:263–71.

Beatty S, Murray IJ, Henson DB, Carden D, Koh H, Boulton ME. Macular pigment and risk for age-related macular degeneration in subjects from a Northern European population. Invest Ophthalmol Vis Sci. 2001;42:439–46.

Berson EL, Rosner B, Sandberg MA, Hayes KC, Nicholson BW, Weigel-DiFranco C, Willett W. A randomized trial of vitamin A and vitamin E supplementation for retinitis pigmentosa. Arch Ophthalmol. 1993;111:761–72.

Bone RA, Landrum JT. Distribution of macular pigment components, zeaxanthin and lutein, in human retina. Methods Enzymol. 1992;213:360–6.

Bonnefont-Rousselot D. Glucose and reactive oxygen species. Curr Opin Clin Nutr Meta Care. 2002;5:561–8.

Boulton M, Rozanowska M, Rozanowski B. Retinal photodamage. J Photochem Photobiol B. 2001;64:144–61.

Bowes C, Li T, Danciger M, Baxter LC, Applebury ML, Farber DB. Retinal degeneration in the rd mouse is caused by a defect in the beta subunit of rod cGMP-phosphodiesterase. Nature. 1990;347:677–80.

Bucheli P, Vidal K, Shen L, Gu Z, Zhang C, Miller LE, Wang J. Goji berry effects on macular characteristics and plasma antioxidant levels. Optom Vis Sci. 2011;88:257–62.

Bhutto I, Lutty G. Understanding age-related macular degeneration (AMD): relationships between the photoreceptor/retinal pigment epithelium/Bruch's membrane/choriocapillaris complex. Mol Aspects Med. 2012;33:295–317.

Caldwell RB, Bartoli M, Behzadian MA, El-Remessy AE, Al-Shabrawey M, Platt DH. Vascular endothelial growth factor and diabetic retinopathy: role of oxidative stress. Curr Drug Targets. 2005;6:511–24.

Caldwell RB, Zhang W, Romero MJ, Caldwell RW. Vascular dysfunction in retinopathy-an emerging role for arginase. Brain Res Bull. 2010;81:303–9.

Carnés JL, De Larramendi CH, Ferrer A, Huertas AJ, López-Matas MA, Pagán JA, Navarro LA, García-Abujeta JL, Vicario S, Peña M. Recently introduced foods as new allergenic sources: sensitisation to Goji berries (*Lycium barbarum*). Food Chem. 2013;137:130–5.

Carpentier S, Knaus M, Suh M. Associations between lutein, zeaxanthin, and age-related macular degeneration: an overview. Crit Rev Food Sci Nutr. 2009;49:313–26.

Castellano G, González-Santander JL, Lara A, Torrens F. Classification of flavonoid compounds by using entropy of information theory. Phytochemistry. 2012a;93:182–91.

Castellano G, Tena J, Torrens F. Classification of phenolic compounds by chemical structural indicators and Its relation to antioxidant properties of posidonia oceanica (L.) Delile. MATCH Commun Math Comput Chem. 2012b;67:231–50.

Castellano G, González-Santander JL, Lara A, Torrens F. Classification of flavonoid compounds by using entropy of information theory. Phytochemistry. 2013;93:182–91.

Chan HC, Chang RC, Koon-Ching Ip A, Chiu K, Yuen WH, Zee SY, So KF. Neuroprotective effects of *Lycium barbarum* Lynn on protecting retinal ganglion cells in an ocular hypertension model of glaucoma. Exp Neurol. 2007;203:269–73.

Chang RCC, So KF. Use of anti-aging herbal medicine, *Lycium barbarum*, against aging-associated diseases. What do we know so far?. Cell Mol Neurobiol. 2008;28:643–52.

Chen Z, Tan BKH, Chan SH. Activation of T lymphocytes by polysaccharide–protein complex from *Lycium barbarum* L. Int Immunopharmacol. 2008;8:1663–71.

Chiu K, Chan HC, Yeung SC, Yuen WH, Zee SY, Chang RC, So KF. Modulation of microglia by wolfberry on the survival of retinal ganglion cells in a rat ocular hypertension model. J Ocul Biol Dis Infor. 2009;2:47–56.

Chiu K, Zhou Y, Yeung SC, Lok CK, Chan OO, Chang RC, So KF, Chiu JF. Up-regulation of crystallins is involved in the neuroprotective effect of wolfberry on survival of retinal ganglion cells in rat ocular hypertension model. J Cell Biochem. 2010;110:311–20.

Chong KP, Rossall S, Atong M. *In vitro* antimicrobial activity and fungitoxicity of syringic acid, caffeic acid and 4-hydroxybenzoic acid in *Ganoderma Boninense*. J Agr Sci. 2010;1:15–20.

Chucair AJ, Rotstein NP, Sangiovanni JP, During A, Chew EY, Politi LE. Lutein and zeaxanthin protect photoreceptors from apoptosis induced by oxidative stress: relation with docosahexaenoic acid. Invest Ophthalmol Vis Sci. 2007;48:5168–77.

Croft KD. The chemistry and biological effects of flavonoids and phenolic acids. Ann. N Y Acad Sci. 1998;854:435–42.

Cui Y, Xu X, Bi H, Zhu Q, Wu J, Xia X, Qiushi Ren, Ho PC. Expression modification of uncoupling proteins and MnSOD in retinal endothelial cells and pericytes induced by high glucose: the role of reactive oxygen species in diabetic retinopathy. Exp Eye Res. 2006;83:807–16.

Cui, BK, Liu S, Lin XJ, Wang J, Li SH, Wang QB, Li SP. Effects of *Lycium barbarum* aqueous and ethanol extracts on high-fat-diet induced oxidative stress in rat liver tissue. Molecules. 2011;16:9116–28.

Dai J, Mumper RJ. Plant phenolics: extraction, analysis and their antioxidant and anticancer properties. Molecules. 2010;15:7313–52.

Ding X, Patel M, Chan CC. Molecular pathology of age-related macular degeneration. Prog Retin Eye Res. 2009;28:1–18.

Dong J Z, Lu DY, Wang Y. Analysis of flavonoids from leaves of cultivated *Lycium barbarum* L. Plant Foods Hum Nutr. 2009;64:199–204.

Du Y, Miller CM, Kern TS. Hyperglycemia increases mitochondrial superoxide in retina and retinal cells. Free Radic Biol Med. 2003;35:1491–9.

Ellis EA, Guberski DL, Somogyi-Mann M, Grant MB. Increased H2O2, vascular endothelial growth factor and receptors in the retina of the BBZ/WOR diabetic rat. Free Radic Biol Med. 2000;28:91–101.

Erlund I. Rewiew of the flavonoids quercetin, hespetin and naringernin: dietary sources, bioactivities, bioavailability and epidemiology. Nutr Res. 2004;24:851–74.

Evans JR. *Ginkgo biloba* extract for age-related macular degeneration. Cochrane Database Syst Rev. 2013;1:CD001775.

Fies P, Dienel A. Ginkgo extract in impaired vision–treatment with special extract EGb 761 of impaired vision due to dry senile macular degeneration. Wien Med Wochenschr. 2002;152:423–6.

Friedman DS, O'Colmain BJ, Munoz B, Tomany SC, McCarty C, de Jong PT, Nemesure B, Mitchell P, Kempen J. Prevalence of age-related macular degeneration in the United States. Arch Ophthalmol. 2004;122:564–72.

Ganea E, Harding JJ. Glutathione-related enzymes and the eye. Curr Eye Res. 2006;31:1–11.

González, E. Evaluación clínica y electrofisiológica de una asociación de suplementos dietéticos con luteína y zeaxantiana en la progesión de la degeneración macular y la calidad de vida en pacientes con Degeneración Macular Asociada a la Edad Atrófica. Thesis Universidad Católica de Valencia. Valencia; 2013.

Halliwell, B. Reactive oxygen species and the central nervous system. J Neurochem. 1992;59:1609–23.

Hammerbacher A, Ralph SG, Bohlmann J, Fenning TM, Gershenzon J, Schmidt A. Biosynthesis of the major tetrahydroxystilbenes in spruce, astringin and isorhapontin, proceeds via resveratrol and is enhanced by fungal infection. Plant Physiol. 2011;157:876–90.

Hartong DT1, Berson EL, Dryja TP. Retinitis pigmentosa. Lancet. 2006;368:1795–809.

He S, Wu B, Pan Y, Jiang L. Chunganenol: an unusual antioxidative resveratrol hexamer from *Vitis chunganensis*. J Org Chem. 2008;73:5233–41.

He M, Pan H, Chang RC, So KF, Brecha NC, Pu M. Activation of the Nrf2/HO-1 antioxidant pathway contributes to the protective effects of *Lycium barbarum* polysaccharides in the rodent retina after ischemia-reperfusion-induced damage. PLoS One. 2014;9:e84800. doi: 10.1371/journal.pone.0084800.

Hozawa A, Jacobs DR, Steffes MW, Gross MD, Steffen LM, Lee D. Circulating carotenoid concentrations and incident hypertension: the coronary artery risk development in young adults (CARDIA) study. J Hypertens. 2009;27:237–42.

Hsu YL, Kuo PL, Lin CC. Acacetin inhibits the proliferation of Hep G2 by blocking cell cycle progression and inducing apoptosis. Biochem Pharmacol. 2004;67:823–29.

Hu CK1, Lee YJ, Colitz CM, Chang CJ, Lin CT. The protective effects of *Lycium barbarum* and *chrysanthemum morifolum* on diabetic retinopathies in rats. Vet Ophthalmol Suppl. 2012;2: 65–71.

Inbaraja BS, Lua H, Hungb CF, Wub WB, Linc C L, Chen BH. Determination of carotenoids and their esters in fruits of *Lycium barbarum Linnaeus* by HPLC–DAD–APCI–MS. J Pharm Biomed Anal. 2008;47:812–18.

Jarrett SG, Boulton ME. Consequences of oxidative stress in age-related macular degeneration. Mol Aspects Med. 2012;33:399–417.

Jarrett SG, Lin H, Godley BF, Boulton ME. Mitochondrial DNA damage and its potential role in retinal degeneration. Prog Retin Eye Res. 2008;27:596–607.

Jin M, Lu Z, Huang M, Wang Y, Wang Y. Sulfated modification and antioxidant activity of exopolysaccahrides produced by Enterobacter cloacae Z0206 Int J Biol Macromol. 2011;48:607–12.

Jin M, Huang Q, Zhao K, Shang P. Biological activities and potential health benefit effects of polysaccharides isolated from *Lycium barbarum L*. Int J Biol Macromol. 2013;54:16–23.

Kadoma Y, Fujisawa S. A comparative study of the radical-scavenging activity of the phenolcarboxylic acids caffeic acid, p-coumaric acid, chlorogenic acid and ferulic acid, with or without 2′-mercaptoethanol, a thiol, using the induction period method. Molecules. 2008;13:2488–99.

Kansanen E, Kuosmanen SM, Leinonen H, Levonen AL. The Keap1-Nrf2 pathway: mechanisms of activation and dysregulation in cancer. Redox Biol. 2013;1:45–9.

Kern TS, Engerman RL. Pharmacological inhibition of diabetic retinopathy: aminoguanidine and aspirin. Diabetes. 2001;50:1636–42.

Khadem S, Marles RJ. Monocyclic phenolic acids; hydroxy- and polyhydroxybenzoic acids: occurrence and recent bioactivity studies. Molecules. 2010;15:7985–8005.

Khandhadia S, Lotery A. Oxidation and age-related macular degeneration: insights from molecular biology. Expert Rev Mol Med. 2010;12:e34.

Kim HP, Kim SY, Lee EJ, Kim YC, Kim YC. Zeaxanthin dipalmitate from *Lycium chinense* has hepatoprotective activity. Res Commun Mol Pathol Pharmacol. 1997;97:301–14.

Kolb H. How the retina works. Am Sci. 2003;91:28–35.

Komeima K, Rogers BS, Lu L, Campochiaro PA. Antioxidants reduce cone cell death in a model of retinitis pigmentosa. Proc Natl Acad Sci USA. 2006;103:11300–5.

Kowluru RA. Diabetic retinopathy: mitochondrial dysfunction and retinal capillary cell death. Antioxid Redox Signal. 2005;7:1581–7.

Kowluru RA, Abbas SN. Diabetes-induced mitochondrial dysfunction in the retina. Invest Ophthalmol Vis Sci. 2003;44:5327–34.

Kukidome D, Nishikawa T, Sonoda K, Imoto K, Fujisawa K, Yano M, Motoshima H, Taguchi T, Matsumura T, Araki E. Activation of AMP-activated protein kinase reduces hyperglycemia-induced mitochondrial reactive oxygen species production and promotes mitochondrial biogenesis in human umbilical vein endothelial cells. Diabetes. 2006;55:120–7.

Lacikova L, Jancova M, Muselik J, Masterova I, Grancai D, Fickova M. Antiproliferative, cytotoxic, antioxidant activity and polyphenols contents in leaves of four staphylea L. species. Molecules. 2009;14:3259–67.

Lebuisson DA, Leroy L, Rigal G. Treatment of senile macular degeneration with *Ginkgo biloba* extract. A preliminary double-blind drug vs. placebo study. Presse Med. 1986;15:1556–8.

Lecleire-Collet A, Tessier LH, Massin P, Forster V, Brasseur G, Sahel JA, Picaud S. Advanced glycation end products can induce glial reaction and neuronal degeneration in retinal explants. Br J Ophthalmol. 2005;89: 1631–3.

Li XN, Song J, Zhang L, LeMaire SA, Hou X, Zhang C, Coselli JS, Chen L, Wang XL, Zhang Y, Shen YH. Activation of the AMPK-FOXO3 pathway reduces fatty acid-induced increase in intracellular reactive oxygen species by upregulating thioredoxin. Diabetes. 2009;58:2246–57.

Li SY, Yang D, Yeung CM, Yu WY, Chang RC, So KF, Wong D, Lo AC. *Lycium barbarum* polysaccharides reduce neuronal damage, blood-retinal barrier disruption and oxidative stress in retinal ischemia/reperfusion injury. PLoS One. 2011;6:e16380. doi: 10.1371/journal.pone.0016380.

Liang FQ, Godley BF. Oxidative stress-induced mitochondrial DNA damage in human retinal pigment epithelial cells: a possible mechanism for RPE aging and age-related macular degeneration. Exp Eye Res. 2003;76:397–403.

Lidebjer C, Leanderson P, Ernerudh J, Jonasson L. Low plasma levels of oxygenated carotenoids in patients with coronary artery disease. Nutr Metab Cardiovasc Dis. 2007;17:448–56.

Lin H, Qian J, Castillo AC, Long B, Keyes KT, Chen G, Ye Y. Effect of miR-23 on oxidant-induced injury in human retinal pigment epithelial cells. Invest Ophthalmol Vis Sci. 2011a;52:6308–14.

Lin NC, Lin JC, Chen SH, Ho CT, Yeh AI. Effect of Goji (*Lycium barbarum*) on expression of genes related to cell survival. J Agric Food Chem. 2011b;59:10088–96.

Lu M, Kuroki M, Amano S, Tolentino M, Keough K, Kim I, Bucala R, Adamis AP. Advanced glycation end products increase retinal vascular endothelial growth factor expression. J Clin Invest. 1998;101:1219–24.

Luo Q, Cai Y, Yan J, Sun M, Corke H. Hypoglycemic and hypolipidemic effects and antioxidant activity of fruit extracts from *Lycium barbarum*. Life Sci. 2004;76:137–49

Madsen-Bouterse SA, Kowluru RA. Oxidative stress and diabetic retinopathy: pathophysiological mechanisms and treatment perspectives. Rev Endocr Metab Disord. 2008;9:315–27.

Manheimer E, Wieland S, Kimbrough E, Cheng K & Berman BM. Evidence from the cochrane collaboration for traditional chinese medicine therapies. J Altern Complement Med. 2009;15:1001–14.

Mao F, Xiao B, Jiang Z, Zhao J, Huang X, Guo J. Anticancer effect of *Lycium barbarum* polysaccharides on colon cancer cells involves G0/G1 phase arrest. Medical Oncology. 2011;28:121–6.

Marcum JL, Mathenia JK, Chan R, Guttmann RP. Oxidation of thiol-proteases in the hippocampus of Alzheimer's disease. Biochem Biophys Res Commun. 2005;334:342–8.

McLaughlin ME, Ehrhart TL, Berson EL, Dryja TP. Mutation spectrum of the gene encoding the beta subunit of rod phosphodiesterase among patients with autosomal recessive retinitis pigmentosa. Proc Natl Acad Sci USA. 1995;92:3249–53.

Mi XS, Chiu K, Van G, Leung JWC, Lo ACY, Chung SK, Chang RCC, So KF. Effect of *Lycium barbarum* polysaccharides on the expression of endothelin-1 and its receptors in an ocular hypertension model of rat glaucoma. Neural Regen Res. 2012a;7:645–51.

Mi XS, Feng Q, Lo AC, Chang RC, Lin B, Chung SK, So KF. Protection of retinal ganglion cells and retinal vasculature by *Lycium barbarum* polysaccharides in a mouse model of acute ocular hypertension. PLoS One. 2012b;4:e45469. doi:10.1371/journal.pone.0045469.

Miranda M, Muriach M, Johnsen S, Bosch-Morell F, Araiz J, Romá J, Romero FJ. Oxidative stress in a model for experimental diabetic retinopathy: treatment with antioxidants. Arch Soc Esp Oftalmol. 2004;79:289–94.

Miranda M, Muriach M, Roma J, Bosch-Morell F, Genovés JM, Barcia J, Araiz J, Díaz-Llospis M, Romero FJ. Oxidative stress in a model of experimental diabetic retinopathy: the utility of peroxinytrite scavengers. Arch Soc Esp Oftalmol. 2006;81:27–32.

Miranda M, Johnson LE, Ahuja S, Ekstrom PA, Romero FJ, van Veen T. Significant photoreceptor rescue by treatment with a combination of antioxidants in an animal model for retinal degeneration. Neuroscience. 2007;145:1120–9.

Miranda M, Arnal E, Ahuja S, Alvarez-Nölting R, López-Pedrajas R, Ekström P, Bosch-Morell F, van Veen T, Romero FJ. Antioxidants rescue photoreceptors in rd1 mice: Relationship with thiol metabolism. Free Radic Biol Med. 2010;48:216–22.

Moeller SM, Parekh N, Tinker L, Ritenbaugh C, Blodi B, Wallace RB, Mares JA. Associations between intermediate age-related macular degeneration and lutein and zeaxanthin in the carotenoids in age-related eye disease study (CAREDS): ancillary study of the women's health initiative. Arch Ophthalmol. 2006;124:1151–62.

Montezuma SR, Sobrin L, Seddon JM. Review of genetics in age related macular degeneration. Semin Ophthalmol. 2007;22:229–40.

Moore TC, Moore JE, Kaji Y, Frizzell N, Usui T, Poulaki V, Campbell IL, Stitt AW, Gardiner TA, Archer DB, Adamis AP. The role of advanced glycation end products in retinal microvascular leukostasis. Invest Ophthalmol Vis Sci. 2003;44:4457–64.

Muriach M, Bosch-Morell F, Alexander G, Blomhoff R, Barcia J, Arnal E, Almansa I, Romero FJ, Miranda M. Lutein effect on retina and hippocampus of diabetic mice. Free Radic Biol Med. 2006;41:979–84.

Natella F, Nardini M, Di Felice M, Scaccini C. Benzoic and cinnamic acid derivatives as antioxidants: structure-activity relation. J Agric Food Chem. 1999;47:1453–9.

Obrosova IG, Minchenko AG, Vasupuram R, White L, Abatan OI, Kumagai AK. Aldose reductase inhibitor fidarestat prevents retinal oxidative stress and vascular endothelial growth factor overexpression in streptozotocin-diabetic rats. Diabetes. 2003;52:864–71.

Obrosova IG, Pacher P, Szabo C, Zsengeller Z, Hirooka H, Stevens MJ. Aldose reductase inhibition counteracts oxidativenitrosative stress and poly(ADPribose) polymerase activation in tissue sites for diabetes complications. Diabetes. 2005;54:234–42.

Omata Y, Saito Y, Yoshida Y, Niki E. Simple assessment of radical scavenging capacity of beverages. J Agric Food Chem. 2008;56:3386–90.

Packer L, Kraemer K, Rimbach G. Molecular aspects of lipoic acid in the prevention of diabetic complications. Nutrition. 2001;17:888–95.

Paquet-Durand F, Hauck SM, van Veen T, Ueffing M, Ekström P. PKG activity causes photoreceptor cell death in two retinitis pigmentosa models. J Neurochem. 2009;108:796–810.

Pavan B, Capuzzo A, Forlani G. High glucose-induced barrier impairment of human retinal pigment epithelium is ameliorated by treatment with Goji berry extracts through modulation of cAMP levels. Exp Eye Res. 2014;120:50–4.

Peng Y, Ma C, Li Y, Leung KS, Jiang ZH, Zhao Z. Quantification of zeaxanthin dipalmitate and total carotenoids in Lycium fruits (*Fructus Lycii*). Plant Foods Hum Nutr. 2005;60:161–4.

Pereira DM, Valentao P, Pereira JA, Andrade PB. Phenolics: from chemistry to biology. Molecules. 2009;14:2202–11.

Pescosolido N, Del Bianco G, de Feo G, Madia F, Risuleo G, Scarsella G. Induced acute ocular hypertension: mode of retinal cell degeneration. Acta Ophthalmol Scand Suppl. 1998;227:20–1.

Pietta P. Flavonoids as antioxidants. J Nat Prod. 2000;63:1035–42.

Puertas FJ, Díaz-Llopis M, Chipont E, Romá J, Raya A, Romero FJ. Glutathione system of human retina: enzymatic conjugation of lipid peroxidation products. Free Radic Biol Med. 1993;14:549–51.

Punzo C, Kornacker K, Cepko CL. Stimulation of the insulin/mTOR pathway delays cone death in a mouse model of retinitis pigmentosa. Nat Neurosci. 2009;12:44–52.

Rosemann D, Heller W, Sandermann H Jr. Biochemical plant responses to ozone. Plant Physiol. 1991;97:1280–6.

Sancho-Pelluz J, Arango-Gonzalez B, Kustermann S, Romero FJ, van Veen T, Zrenner E, Ekström P, Paquet-Durand F. Photoreceptor cell death mechanisms in inherited retinal degeneration. Mol Neurobiol. 2008;38:253–69.

Sanz MM, Johnson LE, Ahuja S, Ekström PA, Romero J, van Veen T. Significant photoreceptor rescue by treatment with a combination of antioxidants in an animal model for retinal degeneration. Neuroscience. 2007;145:1120–9.

Scarsella G, Nebbioso M, Stefanini S, Pescosolido N. Degenerative effects in rat eyes after experimental ocular hypertension. Eur J Histochem. 2012;54:e42. doi:10.4081/ejh.2012.e42.

Schmidt AM, Vianna M, Gerlach M, Brett J, Ryan J, Kao J, Esposito C, Hegarty H, Hurley W, Clauss M, Wang F, Pan YCE, Tsang TC, Stern D. Isolation and characterization of two binding proteins for advanced glycosylation end products from bovine lung which are present on the endothelial cell surface. J Biol Chem. 1992;267:14987–97.

Schmidt AM, Hori O, Chen JX, Li JF, Crandall J, Zhang J, Cao R, Yan SD, Brett J, Stern D. Advanced glycation end products interacting with their endothelial receptor induce expression of vascular cell adhesion molecule-1 (VCAM-1) in cultured human endothelial cells and in mice: a potential mechanism for the accelerated vasculopathy of diabetes. J Clin Invest. 1995;96:1395–403.

Shan X, Zhou J, Ma T, Chai Q. *Lycium barbarum* polysaccharides reduce exercise-induced oxidative stress. Int J Mol Sci. 2011;12:1081–8.

Shen J, Yang X, Dong A, Petters RM, Peng YW, Wong F, Campochiaro PA. Oxidative damage is a potential cause of cone cell death in retinitis pigmentosa. J Cell Physiol. 2005;203:457–64.

Simic A, Manojlovic D, Segan D, Todorovic M. Electrochemical behavior and antioxidant and prooxidant activity of natural phenolics. Molecules. 2007;12:2327–40.

Song MK, Roufogalis BD, Huang TH. Reversal of the caspase-dependent apoptotic cytotoxicity pathway by taurine from *Lycium barbarum* (Goji Berry) in human retinal pigment epithelial cells: potential benefit in diabetic retinopathy. Evid Based Complement Alternat Med. 2012;2012:323784. doi:10.1155/2012/323784.

Spraul CW, Grossniklaus HE. Characteristics of Drusen and Bruch's membrane in postmortem eyes with age-related macular degeneration. Arch Ophthalmol. 1997;115:267–73.

Stitt AW, Curtis TM. Advanced glycation and retinal pathology during diabetes. Pharmacol Rep. 2005;57:156–68.

Sun MH, Pang JH, Chen SL, Han WH, Ho TC, Chen KJ, Kao LY, Lin KK, Tsao YP. Retinal protection from acute glaucoma-induced ischemia-reperfusion injury through pharmacologic induction of heme oxygenase-1. Invest Ophthalmol Vis Sci. 2010;51:4798–808.

Swaroop A, Chew EY, Rickman CB, Abecasis GR. Unraveling a multifactorial late-onset disease: from genetic susceptibility to disease mechanisms for age-related macular degeneration. Annu Rev Genomics Hum Genet. 2009;10:19–43.

Szwajgier D, Pielecki J, Targon'ski Z. Antioxidant activities of cinnamic and benzoic acid derivatives. Acta Sci Pol Technol Aliment. 2005;4:129–42.

Tan JS, Wang JJ, Flood V, Rochtchina E, Smith W, Mitchell P. Dietary antioxidants and the long-term incidence of age-related macular degeneration: the blue mountains eye study. Ophthalmology 2008;115:334–41.

Tang L, Zhang Y, Jiang Y, Willard L, Ortiz E, Wark L, Medeiros D, Lin D. Dietary wolfberry ameliorates retinal structure abnormalities in db/db mice at the early stage of diabetes. Exp Biol Med (Maywood). 2011;236:1051–63.

Tanito M, Nishiyama A, Tanaka T, Masutani H, Nakamura H, Yodoi J, Ohira A. Change of redox status and modulation by thiol replenishment in retinal photooxidative damage. Invest. Ophthalmol Visual Sci. 2002;43:2392–400.

Ting AY, Lee TK, MacDonald IM. Genetics of age-related macular degeneration. Curr Opin Ophthalmol. 2009;20:369–76.

Vidal K, Bucheli P, Gao Q, Moulin J, Shen LS, Wang J, Blum S, Benyacoub J. Immunomodulatory effects of dietary supplementation with a milk-based wolfberry formulation in healthy elderly: a randomized, double-blind, placebo-controlled trial. Rejuvenation Res. 2012;15:89–97.

Vlachantoni D, Bramall AN, Murphy MP, Taylor RW, Shu X, Tulloch B, Van Veen T, Turnbull DM, McInnes RR, Wright AF. Evidence of severe mitochondrial oxidative stress and a protective effect of low oxygen in mouse models of inherited photoreceptor degeneration. Hum Mol Genet. 2011;2:322–35.

Wang CC, Chang SC, Stephen Inbaraj B, Chen BH. Isolation of carotenoids, flavonoids and polysaccharides from *Lycium barbarum L.* and evaluation of antioxidant activity. Food Chem. 2010;120:184–92.

Warboys CM, Toh HB, Fraser PA. Role of NADPH oxidase in retinal microvascular permeability increase by RAGE activation. Br J Ophthalmol. 2005;89:1631–3.

Wautier JL, Zoukourian C, Chappey O, Wautier MP, Guillausseau PJ, Cao R, Hori O, Stern D, Schmidt AM. Receptor-mediated endothelial cell dysfunction in diabetic vasculopathy: soluble receptor for advanced glycation end products blocks hyperpermeability in diabetic rats. J Clin Invest. 1996;97:238–43.

Webster RP, Gawde MD, Bhattacharya RK. Protective effect of rutin, a flavonol glycoside, on the carcinogen-induced DNA damage and repair enzymes in rats. Cancer Lett. 1996;109:185–91.

Wellard J, Lee D, Valter K, Stone J. Photoreceptors in the rat retina are specifically vulnerable to both hypoxia and hyperoxia. Vis Neurosci. 2005;22:501–7.

Winkler BS, Boulton ME, Gottsch JD, Sternberg P. Oxidative damage and age-related macular degeneration. Mol vis. 1999; 5:32.

Wu HT, He XJ, Hong YK, Ma T, Xu YP, Li HH. Chemical characterization of *Lycium barbarum* polysaccharides and its inhibition against liver oxidative injury of high-fat mice. Int J Biol Macromol. 2010;46:540–3.

Wu WB, Hung DK, Chang FW, Ong ET, Chen BH. Anti-inflammatory and anti-angiogenic effects of flavonoids isolated from *Lycium barbarum Linnaeus* on human umbilical vein endothelial cells. Food Funct. 2012;3:1068–81.

Xin YF, Wan LL, Peng JL, Guo C. Alleviation of the acute doxorubicin-induced cardiotoxicity by *Lycium barbarum* polysaccharides through the suppression of oxidative stress. Food Chem Toxicol. 2011;49:259–64.

Yamagishi S, Ueda S, Matsui T, Nakamura K, Okuda S. Role of advanced glycation end products (AGEs) and oxidative stress in diabetic retinopathy. Curr Pharm Des. 2008;14:962–8.

Yamashima T. Ca2 + -dependent proteases in ischemic neuronal death: a conserved 'calpain-cathepsin cascade' from nematodes to primates. Cell Calcium. 2004;36:285–93.

Yang M, Gao N. protective effect of *Lycium barbarum* polysaccharide on retinal ganglion cells *in vitro*. Int J Ophthalmol. 2011;4:377–9.

Yang CS, Landau JM, Huang M, Newmark HL. Inhibition of carcinogenesis by dietarypolyphenolic compounds. Annu Rev Nutr. 2001;21:381–406.

Yang D, Elner SG, Bian ZM, Till GO, Petty HR, Elner VM. Pro-inflammatory cytokines increase reactive oxygen species through mitochondria and NADPH oxidase in cultured RPE cells. Exp Eye Res. 2007;85:462–72.

Yu H, Wark L, Ji H, Willard L, Jaing Y, Han J, He H, Ortiz E, Zhang Y, Medeiros DM, Lin D. Dietary wolfberry upregulates carotenoid metabolic genes and enhances mitochondrial biogenesis in the retina of db/db diabetic mice. Mol Nutr Food Res. 2013;57:1158–69.

Zarbin MA. Current concepts in the pathogenesis of age-related macular degeneration. Arch Ophthalmol. 2004;122:598–614.

Zhang M, Chen H, Huang J, Li Z, Zhu C, Zhang S. Efect of *Lycium barbarum* polysaccharide on human hepatoma QGY7703 cells: inhibition of proliferation and induction of apoptosis. Life Sci. 2005;76:2115–24.

Zhang W, Zhang X, Wang H, Guo X, Li H, Wang Y, Xu X, Tan L, Mashek MT, Zhang C, Chen Y, Mashek DG, Foretz M, Zhu C, Zhou H, Liu X, Viollet B, Wu C, Huo Y. AMP-activated protein kinase α1 protects against diet-induced insulin resistance and obesity. Diabetes. 2012;61:3114–25.

Zhao Z, Luo Y, Li G, Zhu L, Wang Y, Zhang X. Thoracic aorta vasoreactivity in rats under exhaustive exercise: effects of *Lycium barbarum* polysaccharides supplementation. J Int Soc Sports Nutr. 2013;10:47.

Zhong Y, Shahidi F, Naczk M. Phytochemicals and health benefits of goji berries. In: Dried Fruits: Phytochemicals and Health Effects, 1st edn. New York: Wiley-Blackwell; 2013. pp. 133–44

Ziaei A, Schmedt T, Chen Y, Jurkunas UV. Sulforaphane decreases endothelial cell apoptosis in fuchs endothelial corneal dystrophy: a novel treatment. Invest Ophthalmol Vis Sci. 2013;54:6724–34.

Chapter 12
Allergenic Sensitisation Mediated by Wolfberry

Jerónimo Carnés, Carlos H. de Larramendi, María Angeles López-Matas, Angel Ferrer and Julio Huertas

Abstract Food allergy has increased dramatically in the last 20 years mainly in children and is sometimes mediated by cross-reactivity with other allergens, including pollens or other foods, sometimes by primary sensitisation after the ingestion of the offending foods. On the other hand, globalisation, sophisticated preservation technology or rapid transport methods, as well as the interest for new tastes, food varieties and eating habits in the population are the main causes for the introduction of new foods which immediately become potential allergenic sources.

In this way, the consumption of wolfberries has been generalised in Western countries during the last years. Until now only a few cases of allergic reactions have been reported in scientific papers since 2011, but the appearance of new cases is getting more frequent progressively. Allergenic clinical symptoms oscillate between mild reactions, mainly located in the oral cavity (oral allergic syndrome), to generalised symptoms, or even anaphylactic reactions with urticaria, dyspnoea, oedemaa and/or rhinitis. A pioneer multi-centre observational prospective study, conducted with 566 individuals residing on the Spanish Mediterranean coast, estimated the prevalence of allergic sensitisation to wolfberries at 5.8%. However, only 18% of the total population reported consumption. These results suggested that most of the allergic reactions could be mediated by cross-reactivity with other foods habitually consumed by individuals. In spite of the variety and severity of symptoms and the high rate of prevalence sensitisation, allergy to wolfberries is not very frequent.

Although the number of proteins present in wolfberry allergenic extracts is very numerous, its allergenic composition, determined by using a specific pool of sera

J. Carnés (✉) · M. A. López-Matas
R&D Department, Laboratorios LETI S.L., Calle del Sol, 5, Tres Cantos, 28760 Madrid, Spain
e-mail: jcarnes@leti.com

C. H. de Larramendi
Allergy Section, Hospital Marina Baixa, Villajoyosa and Centro de Especialidades Foietes, Benidorm, Alicante, Spain

A. Ferrer
Allergy Unit, Hospital General Universitario de Elche, Alicante, Spain

J. Huertas
Allergy Section, Complejo Hospitalario Universitario de Cartagena, Murcia, Spain

© Springer Science+Business Media Dordrecht 2015
R. C-C. Chang, K-F. So (eds.), *Lycium Barbarum and Human Health,*
DOI 10.1007/978-94-017-9658-3_12

from allergic individuals, is formed by eight different proteins with capacity to bind specific immunoglobulin E (IgE). These allergens range in a molecular weight between approximately 7 and 70 kDa. Until now, an approximately 7-kDa allergen has been identified as the most relevant, with a frequency of recognition by sensitised individuals higher than 80 %. Proteomics studies showed that this allergen, tentatively named Lyc bar 3, corresponds to the group of non-specific lipid transfer proteins (nsLTP), a family of pan-allergens widely spread in the plant kingdom and responsible for a large number of allergic sensitisations.

Based on in vitro studies, IgE cross-reactivity between wolfberries and other members of the *Solanaceae* family has been clearly established. This concept demonstrates a similarity not only in the allergen composition but also in the protein or IgE epitopes structure of different members of the family. Therefore, allergy to wolfberries in countries where it has been recently introduced is related with high cross-reactivity with other frequently consumed foods and explains the high rate of prevalence and the low frequency of clinical symptoms.

Keywords Food allergy · New food · Concomitant sensitisation · Goji berries · Wolfberry · Allergens · Cross-reactivity · Panallergens · nsLTP

12.1 Introduction

Allergy is a major health problem that is becoming more common. During the last 20 years, the number of people suffering from allergies has increased rapidly. Although, in general terms, allergy is not a common life-threatening disease, fatal reactions are reported every year, especially related to food allergy. In this context, the psychological impact of food allergy in children, adolescents and their families is a major concern because food allergy can affect social life, habits, diets and behaviour.

Globalisation enables the continuous introduction of new foods and flavours into people's daily diet across the world. This leads to a new challenge in allergy prevention. The consumption of wolfberries (goji berries) in Western countries is a clear example of this situation. From an allergic perspective, food or other allergenic sources are not dangerous for humans per se. The immune system of susceptible individuals converts proteins into allergens through the incorrect processing of the ingested proteins and their transformation into substances that are aggressive for the human body. Some of these incorrectly processed proteins become allergenic molecules with the capacity to induce primary sensitisation or adverse reactions mediated by cross-reactivity with primary sensitizers.

The introduction of new diagnostic techniques, such as molecular diagnosis, is greatly helping to elucidate the allergenic profiles of individuals and the responsible allergens, and is providing a deeper knowledge of the associations between different allergenic sources and individuals. However, we are still very far from being able to predict and prevent food allergies.

12.2 Food Allergy

Food allergy can be described as an adverse immune reaction induced by the ingestion of dietary or food antigens present in food.

Ingested food is processed in the gastrointestinal or alimentary tract. This has two important functions: the first is nutritional, involving processing and transforming the ingested food to obtain all the nutrients and oligoelements needed to survive, and the second is related to immunologic tolerance. The gastrointestinal tract processes ingested food and many other different potentially harmful antigenic substances, which have to be neutralised or blocked before absorption. Large amounts of immunologically active food proteins penetrate routinely into the bloodstream, but immunological tolerance prevents the occurrence of adverse reactions. A failure in this tolerance process will result in an allergic reaction against the culprit ingested food. Immunological tolerance is crucial for protecting individuals. While self-antigen reactions are prevented by the stimulation of T-regulatory lymphocytes and T-cell receptor (TCR) in the thymus, tolerance to food antigens is not fully understood and it involves the activation of a number of different types of T-regulatory cells (Himmel et al. 2012; Rescigno 2011).

Although different classifications of the concept food allergy have been published, from an immunological point of view, it can be divided into three groups based on the involvement of immunoglobulin E (IgE) in the adverse reaction:

- IgE-mediated reactions. These allergic reactions are mediated by specific IgEs, namely when these immunoglobulins, formed during the sensitisation process, initiate the degranulation of inflammatory mediators. Allergenic sensitisation starts when a protein from food is captured by dendritic cells. These dendritic cells process and present the antigens (termed as allergens) and stimulate a B cell response responsible for the production of IgE antibodies. Circulating specific IgEs produced by B cells are recognised by high-affinity IgE receptors present on the surface of effector cells, including basophils and mast cells. After a second or successive exposure to the allergen, specific IgEs initiate a cascade of reactions resulting in the release of inflammatory mediators such as histamine and tryptase, as well as cytokines and other immunological mediators, which leads to an adverse reaction that may range from mild to moderate or severe reactions, anaphylaxis or even death.
- Non-IgE-mediated reactions. They consist of a heterogeneous group and pathogenic mechanisms that are not always well understood. As the name indicates, these adverse reactions are not initiated by IgEs. The most common forms are celiac disease, dietary protein enterocolitis syndrome and dietary protein enteropathy syndrome. Enterocolitis and enteropathy syndromes are typically provoked in children by cow's milk and are characterised by an increase in lymphocytes, eosinophils and mast cells. Celiac disease is probably the most well-known pathology because it affects people of all ages and is associated with profound villous atrophy and extensive cellular infiltrate. This disease has a complex pathogenic mechanism in which gliadin is essential and induces the production of auto-antibodies against intestinal transglutaminase.

• A mixture of IgE- and non-IgE-mediated reactions. This classification includes a group of diseases characterised by the presence of eosinophilic infiltration with the minor presence of other inflammatory cells. The infiltration can affect any specific region of the gastrointestinal tract, such as eosinophilic esophagitis, which is focalised in the oesophagus, or it can be generalised (eosinophilic gastroenteritis).

12.3 Food Allergens

Thousands of different food proteins/antigens are ingested in the diet. However, only a small number of these proteins have been identified as allergens. Allergens are defined as proteins or glycoproteins with a relatively small molecular size, which normally range from 5–7 to 60–70 kDa. They are usually soluble in aqueous solutions and are able to stimulate an IgE response. Food allergens have additional properties such as resistance to digestion and stability during food processing. It is important to note that, in some cases, cooking or heating can modify the structure of non-allergenic proteins and transform them into allergens (Carnés et al. 2007).

Food allergens can be divided into plant- or animal-derived allergens. Almost 200 food allergens have been identified to date (http://www.allergen.org/), the most common being those derived from animals such as eggs, milk, seafood and fish, as well as fruits, vegetables, nuts and seeds. However, the incidence or importance of the allergenic sources depends primarily on the characteristics and allergenic profiles of individuals and populations (Beitia et al. 2014; Tripodi et al. 2012).

The most relevant plant food allergens are included in just a small number of established families, based on their biological function. More than 70 % of plant-food-derived allergens are contained in only four of these superfamilies (Table 12.1).

The prolamin superfamily comprises the largest number of allergens. Three families of allergens are included within the prolamin superfamily: non-specific lipid transfer proteins (nsLTP), typically found in fruits and vegetables; seed storage albumins, found in seeds and nuts; and amylase/trypsin inhibitors, present in cereals. As a result of their biochemical composition, they are all resistant to digestion and thermal denaturation, which gives them high-potential allergenicity.

The other two most important superfamilies are cupins and profilins. The cupin superfamily comprises heat stable and highly allergenic allergens. The majority are storage proteins present in nuts. The profilins constitute a very conservative family of proteins, which confers them a high degree of cross-reactivity among different species. Profilins are responsible for tree pollinosis-associated food syndrome. Finally, the superfamily of the allergens homologous to Bet v 1 is the fourth superfamily. The food allergic reactions are primarily explained by IgE antibodies to Bet v 1, which cross-react with Bet v 1 homologous proteins. These allergens are mainly localized in the pulp of the fruit. Heat treatment of these foods destroys the native three-dimensional molecular structure but does not affect linear peptides, important for the late phase cellular reaction.

Table 12.1 Classification of plant food allergens indicating their biological function and common allergens

Classification of plant food allergens			
Superfamily	Family	Biological function	Common allergens
Prolamin	2S albumins	Seed storage	Wheat, peanut (Ara h 2, Ara h 6, Ara h 7), walnut (Jug r 1)
	nsLTP	Defensin related	Peach (Pru p 3), hazelnut (Cor a 8), tomato (Sola l 3), apple (Mal d 3), goji berries (Lyc bar 3)
	α-Amylase/tryp-sin inhibitors	Protection against degrada-tion and pathogens	Wheat (Hor v 1)
Cupin	7S albumins (vicilin)	Seed storage	Buckwheat (Fag e 3), peanut (Ara h 1)
	11S albumins (legumin)	Seed storage	Peanut (Ara h 3)
Profilins		Structural proteins	Peanut (Ara h 5), hazelnut (Cor a 2), apple (Mal d 4), wheat (Tri a 12)
Homologous to Bet v 1		Pathogenesis related pro-teins (PR-10)	Carrot (Dau c 1)

12.4   Cross-Reactivity

The term cross-reactivity is associated with the allergic clinical symptoms that individuals may experience after the consumption of/contact with some types of foods or pollens, even though they may have never previously been in contact with them. In contrast, the term co-sensitisation means the actual sensitisation to different allergenic sources with the production of specific antibodies to each of them.

From an allergenic point of view, cross-reactivity can be defined as the capacity of allergic individuals to react against allergenic sources to which they are not sensitised because they have not been exposed to them or have not developed specific antibodies against them. In this case, the components of the immunological system of patients (specific IgE) recognise the allergenic components (allergens) present in the offending food but with the particularity that these allergens are recognised only as a consequence of structural similarities with the allergen to which patients are actually sensitised. This recognition can occur due to the presence of common proteins/allergens in different organisms, known as panallergens, or as a consequence of similarities in specific regions of the proteins.

The term panallergen is used to define proteins/allergens that are widely distributed in different organisms, that have a common biological function and that share a somewhat common structure. nsLTPs and profilins are the most studied group of panallergens. Both are present in all plant kingdom organisms and are responsible for a large number of sensitisations, either by co-sensitisation or cross-reactivity.

The cross-reactivity between profilins from different organisms has been well established (Radauer et al. 2006). However, cross-reactivity between LTPs remains unclear. In a recently published study, the authors demonstrated that, although all LTPs have the same biological function, amino acid sequence varies significantly among organisms (Morales et al. 2014). Immunological studies performed with serum samples from animals and human patients demonstrated only partial cross-reactivity and suggested that, not only the proteins, but also the matrix in which proteins are contained could play a role in allergic symptoms (Schulten et al. 2011) because of its ability to act as an adjuvant for immune response. In this sense, authors suggested that allergic reactions could be related to primary sensitisation and not only to cross-reactivity.

Recognition as a consequence of similarities in specific regions of the proteins is less common and is mediated by proteins which share specific similar structures that can be recognised by common antibodies. This is the typical case of cross-reactivity known as "Bet v 1 Family". In these patients, primary sensitisation seems to be produced by inhaling pollen proteins. Individuals experience symptoms when they ingest Bet v 1 homologous allergens such as carrots or apples (Bohle 2007).

12.5 Prevalence of Food Allergy

Food allergy appears to be a major and increasing problem, especially in Western countries. A dramatic increase in the number of affected populations has been observed in the last 20 years and the percentage of patients with some allergic symptoms increases year-on-year.

In recent years, globalisation, population interest in new flavours, rapid transportation and improvement of food preservation conditions have enabled new foods and flavours to spread throughout the world. Diet diversification and increasing demand for better quality and labour-saving products have increased the imports of high-value and processed food products in developed countries. As a result, individuals experience new and greater access to new allergenic sources to which they had never before been exposed. This may modify the sensitisation pattern of individuals, either because of new allergenic sensitisation provoked by the allergenic capacity of new foods, or individuals may react to the new food by cross-reactivity. The addition of kiwi to the Western diet at the end of the 1970s is a clear example of new sensitisation. Nowadays, kiwi is recognised as a potent allergenic source, and actinidin, one of its major allergens, was previously not habitually consumed in the Western diet. Moreover, goji berries are an example of sensitisation to new allergenic sources by cross-reactivity, in this case mediated by LTPs.

There is no clear consensus regarding the prevalence of food allergy in the general population, although different studies concur that there is a higher prevalence in children. In general terms, it has been estimated that 2 % of the adult population and 6–8 % of children suffer from food allergy. Although there is no clear consensus, the prevalence is probably underestimated because of a considerable under-diagnosed population. Moreover, these values are highly variable according to country. Western countries probably have the highest rates of food allergy. According to the European Academy of Allergy and Clinical Immunology (EAACI) guidelines (Muraro et al. 2014a; Nwaru et al. 2014), more than 17 million individuals suffer from food allergies in Europe. Food sensitisation, understood as patients with specific IgE but with no clinical symptoms, may double or triple this figure.

Double-blind, placebo-controlled food challenge (DBPCFC) is recognised as the most effective tool for allergy diagnosis. However, this diagnostic method has its flaws and is not always a routine practice mainly for methodological reasons. Therefore, actual prevalence of food allergy cannot currently be confirmed in the general population and further studies are required.

Factors such as age, sex, race, country of origin and residence, family or personal history of atopy or other concomitant allergic diseases play an important role in the development of food allergies. All these factors, as well as food consuming habits, may affect the prevalence of food allergy in different populations and explain different prevalence rates or sensitisation to different food allergens among patients living in different areas.

12.6 Clinical Management

Food allergy is associated with psychological distress in patients, especially in children, adolescents and their parents. It also causes anxiety and depression (Knibb and Semper 2013; Bacal 2013) in families as a result of the risk associated with this pathology (Muraro et al. 2014b). Undoubtedly, the most life-threatening effect of food allergy is anaphylaxis, understood to be a generalised or systemic hypersensitivity reaction mediated by the ingestion of food allergens which, on occasions, may be fatal for patients. In the last few years, the number of episodes of anaphylaxis induced by the food ingestion has increased dramatically, especially in developed countries. Anaphylaxis treatment guidelines recommend that at-risk patients carry adrenaline auto-injectors with them at all times (Song et al. 2014).

Pharmacological treatment is useful only for the symptomatic treatment of acute symptoms. Its use is recommended to alleviate mild-to-moderate symptoms and, until now, adrenaline is the only effective treatment for life-threatening episodes (Arnold et al. 2011).

Until now, dietary avoidance has been recognised as the only effective treatment for food allergy. However, long-term avoidance has to be carefully monitored because it can result in nutritional problems and affects patient quality of life.

Specific immunotherapy with food allergens is currently being investigated, although it is not yet suitable for routine practice. The capacity of this kind of immunotherapy to restore permanent tolerance to food has not yet been conclusively demonstrated. Three different approaches are under study, including subcutaneous immunotherapy, sublingual immunotherapy and induction of oral tolerance, also called oral desensitisation. Preliminary results obtained with subcutaneous immunotherapy suggest that this could be an effective alternative. Other studies with sublingual immunotherapy were associated with improved tolerance and the reduction of symptoms (Nowak-Wegrzyn and Albin 2014). Oral desensitisation is also one of the most frequently studied alternatives. Published results showed efficacy on many occasions, but there is no consensus regarding treatment, dose, allergens, time of exposure, long-term studies, etc. (Pajno et al. 2014).

12.7   Goji Berries Consumption-Mediated Allergy

Goji berries, considered both a fruit and a herb, have been part of the common diet in some Asian regions, such as China, Tibet and Mongolia, for more than 3000 years. In some of these countries, they are considered to be a true treasure and their properties almost have a mystical value. They are consumed daily even at high doses sometimes as a medicinal herb to treat diseases due to their healthy properties, or other times in search of "the source of eternal youth", as part of the normal diet. However, to date no well-documented data related to sensitisation or allergic reactions have been reported by countries that were the first consumers, although, urticaria and rashes after goji berries' ingestion has been described.

In Europe, according to the European Union Novel Food Regulation (http://ec.europa.eu/food/food/biotechnology/novelfood/index_en.htm), a food is judged to be novel if it has not been consumed in significant quantities in Europe prior to May 1997. When the consumption of goji berries started to spread throughout European countries, regulatory agencies investigated their previous consumption. Initially, authorities did not find a history of significant consumption of goji berries in Europe before 1997. However, in 2007, the UK Food Standards Agency determined that, in fact, there was a history of significant consumption of the fruit in Europe before 1997, and goji berries were removed from the Novel Foods list.

In the USA and Canada, farmers began cultivating goji berries during the first decade of the twenty-first century on a commercial scale.

Goji berries can therefore be considered as a food recently introduced into the European and North American diet. No cases of allergic reactions had been reported in the scientific literature until recently. Additionally, some occasional reactions such as urticaria-like or papular rashes have been linked to goji berry consumption and published in Chinese medicine textbooks, although they cannot necessarily be classified as allergic reactions (Bensky et al. 2004). Since 1990, more than 3000 adverse reactions have been described in China with more than 210 involving herbs.

The first two cases of anaphylaxis following the ingestion of goji berries that were investigated in depth on a worldwide scale were reported in Spain in 2011, and five additional cases were also reported in Spain in 2012. The consumption of goji berries was the common factor in each case, but the severity of symptoms, origin of patients and concomitant allergies differed. All of them were associated with preliminary sensitisation to LTPs. Table 12.2 details the eight clinical cases described in the literature.

Table 12.2 Clinical characteristics of all goji berries allergic patients published in the literature

Clinical cases with goji berry allergy

Author	Patient	Symptoms	Specific IgE (kilo units of allergen specific-IgE per liter)	Concomitant sensitisation associated to goji berries
Monzon et al. (2011)	27-year-old woman	Grade II anaphylaxis. Acute generalised urticaria, lip oedema, dyspnoea and acute rhinitis	1.38	Food: peach, tomato, green pepper Panallergens: LTP
	13-year-old girl	Generalised urticaria, severe pruritus and skin lesions (hives), angioedema and dysphagia	16.9	Food: peach, kiwi, almond, peanut, hazelnut, chestnut, rice, tomato
				Panallergens: LTP
				Pollens: *Chenopodium album*, *Ambrosia elatior*, *Platanus hybrida*, *Cupressus arizonica*
Larramendi et al. (2012)	40-year-old man	Facial angioedema with dyspnoea	0.78	Food: peach
	31-year-old man	Pharyngeal itching	2.87	Food: peach Panallergens: LTP
	30-year-old woman	Labial angioedema and perioral skin rash	0.58	Food: peach Panallergens: LTP
	36-year-old man	Itching in the mouth, ears and axilla	3.62	Food: peach Panallergens: LTP
	42-year-old woman	Severe generalised itching	0.37	Panallergens: polcalcin
Carnés et al. (2013)	65-year-old man	Oral allergy syndrome	No determined	Mono-sensitised to goji berries
	21-year-old woman	Occasional urticarial and oral allergy syndrome	0.6	Food: tomato Panallergens: LTP
	43-year-old woman	Oral allergy syndrome and diarrhoea	No determined	Mites and cockroach
Gámez et al. (2013)	40-year-old woman	Immediate pharyngeal pruritus	No determined	Foods: fruits, king prawn, nuts, lettuce, escarole and soy Pollens: grasses

IgE immunoglobulin E, *LTP* lipid transfer protein

The first case was a 27-year-old woman who developed grade II anaphylaxis 1 h after goji berry consumption, accompanied by acute generalised urticaria on the hands, palms and soles, lip oedema, dyspnoea and acute rhinitis, and required treatment with adrenaline. The second case was a 13-year-old girl who presented with generalised urticaria, severe pruritus and skin lesions (hives), angioedema and dysphagia after goji berry consumption.

    In 2012, five new cases were reported. In this case, patients showed different severity and manifestations following goji berry consumption. The first patient was a 40-year-old man who complained of facial angioedema with dyspnoea (requiring epinephrine) while eating goji berries for the first time (30–40 berries). The second case was a 31-year-old man who reported pharyngeal itching lasting 30–60 min on 10–12 occasions, immediately after eating 10–15 goji berries and with symptoms increasing in intensity after each exposure. The third case was a 30-year-old woman who reported labial angioedema and perioral skin rash immediately after eating a new pack of 20 goji berries. The patient had previously tolerated a whole pack of goji berries. The fourth case was a 36-year-old man with a history of allergic rhinoconjunctivitis and urticaria due to peanuts who reported itching in the mouth, ears and axilla lasting 10 min immediately after eating a single goji berry. Finally, the fifth case was a 42-year-old woman who reported a 2-month history of severe generalised itching that resolved after avoiding goji berries. She was consuming approximately 20–30 goji berries daily for several months.

    Additionally, in a different study designed to evaluate the prevalence of sensitisation to goji berries in 2012, three subjects reported different symptoms following the consumption of goji berries on several occasions. The three individuals reported oral allergy syndrome, one of them occasional and urticarial and one also reported diarrhoea following consumption (Carnés et al. 2013). The last case published to date (2013) involves a 40-year-old woman who presented with immediate pharyngeal pruritus following goji berry ingestion (Gámez et al. 2013).

    Prior to that, two cases had been reported as communications to Congresses. Although the studies lack in-depth investigation, they described a 29-year-old woman who presented with oropharyngeal itching, wheezing, dyspnoea, generalised urticaria and dizziness without loss of consciousness 30 min after the first intake of goji berries (Cadavid Moreno et al. 2010). The second study consisted of a 28-year-old woman reporting an episode of acute urticaria within 30 min of eating goji berries, having started to consume them regularly 1 week before (Guspi and Baltasar 2011). Both reported previous reactions to nuts, peaches and other fruits.

12.8 Sensitisation to Goji Berries and Prevalence

A pioneering multi-centre observational prospective study was conducted in southeast Spain with the primary objective of determining the prevalence of sensitisation to goji berries in the general population without considering whether or not the individuals included had been in contact with the fruit. Patients attending allergy departments for the first time, for respiratory and/or cutaneous symptoms, and/or suspicion of plant food allergy, and with a clinical indication for the performance of skin prick test to common allergens, were included. Five hundred and sixty-six (566) individuals (220 male, 38.9%; 346 females, 61.1%) with a mean age of 33.4 ± 15.2 years, were skin prick tested with a goji berry allergenic extract. Almost 40% of individuals were familiar with and had occasionally tried goji berries and

15 % reported frequent consumption. Although only three individuals (0.5 %) of the all those included reported mild symptoms after goji berries consumption, 5.8 % of the total population in the study showed sensitisation and specific IgE to goji berry extract. That means that 33 individuals out of the total study population had a positive skin prick test to goji berries. The analysis of the specific IgE in serum of 24 of these individuals, for the determination of in vitro sensitisation, showed that 54.2 % had positive specific IgE to goji berries ranging between 0.35 and 26.5 kUA/l,[1] with a mean value of 5.2 ± 8.6 kilo units of allergen specific-IgE per liter. As expected and as occurs with other allergenic sources, the percentage of sensitisation (positive specific IgE) is much higher than the percentage of the population with allergic symptoms. However, sensitised individuals could potentially develop allergic symptoms in the future.

On the other hand, all the individuals included in the previously mentioned three studies showed positive sensitisation to goji berries by skin prick test as well as specific IgE in serum. It is important to note that all of them developed symptoms immediately after the ingestion of goji berries and had positive skin prick tests and positive specific IgE in serum. These results corroborate that the presence of allergic symptoms correlate with positive sensitisation, thereby confirming goji berries as an allergenic source. Therefore, allergenic reactions (mild or severe) associated with the consumption of goji berries are linked to a positive specific IgE, which confirms that this kind of allergic reaction can be included in a typical IgE-mediated allergy.

Interestingly, the observed prevalence of sensitisation to goji berries was very similar to that observed with other fruits frequently consumed by the Mediterranean population (Larramendi et al. 2008). However, these results highlight that the introduction of new foods into the diet may become responsible for new allergic sensitisations, sometimes due to cross-reactivity with known allergens and sometimes due to primary sensitisation to new proteins.

The unanswered question now is whether or not goji berries are highly potent allergenic sources with their own identity and allergens capable of inducing allergic sensitisation, or whether, on the contrary, they share common allergens with other endemic foods that form part of the common diet of the population. If the latter is true, the question is whether or not sensitisation is mediated by concomitant sensitisation to other fruits.

12.9   Sensitisation Profile and Concomitant Allergies of Individuals Allergic to Goji Berries

Two, three or more different sources of allergenic sensitisation are very common in allergy and it is unusual to observe mono-sensitised individuals, especially in food sensitisation cases and in some regions of Western countries, including Mediterranean countries. In general, patients who suffer from food sensitisation usually

---

[1] IgE≥0.35 is considered positive.

show concomitant sensitisation to other foods or pollen, in most cases, mediated by cross-reactivity. The key issue is to be able to diagnose which of the allergenic sources is responsible for the primary sensitisation in mono-sensitised individuals, or whether poly-sensitised individuals are co-sensitised or cross-sensitised and to define the primary sensitiser.

This is one of the key elements to understand whether newly introduced allergens may be allergic inducers on their own or whether the induced allergic symptoms are a consequence of cross-reactivity.

According to the data published in the scientific literature, goji berry sensitisation always appears to be linked to different concomitant sensitisations and, in some cases, these sensitisations show a statistically significant correlation. In the clinical cases described by Monzón (Monzón Ballarín et al. 2011), the individuals reported severe clinical symptoms after the ingestion of different fruits, vegetables and nuts, and mild symptoms, mainly oral allergy syndrome with other foods. This observation was also confirmed by the cases described by Larramendi (Larramendi et al. 2012). In this study, all patients showed sensitisation to peach, in addition to other foods. The study published by Carnés et al. (2013) showed a statistically positive correlation between patients sensitised to goji berries and other foods such as nuts, tomatoes or peaches. What is more, they were sensitised to purified LTP and polcalcin and not to profilin. In all cases, the results suggested that the presence of common allergens in all the allergenic sources may be responsible for goji berry sensitisation and as a result, cross-reactivity may explain allergy to goji berries. This does not mean that goji berries are not capable of being a primary sensitiser. In the Mediterranean population, where allergy to fruits is very common (especially fruits from the *Rosaceae* taxonomical family that are consumed daily and are responsible for a large number of allergic sensitisations), these fruits, particularly peaches, are the primary sensitisers (Boyano-Martínez et al. 2013). The results of this study and the low number of symptomatic patients described explain the characteristics of the population and their exposure to goji berries. According to the study, 18 % of the patients included in the study had tried goji berries, demonstrating that goji berries are not widely consumed and suggesting a potential explanation for the low number of allergic individuals. Nearly 25 % of the subjects with positive sensitisation had tried goji berries and nearly 10 % of them were sensitised, suggesting that the allergenic potential of these fruits is high. These results suggest that the risk of developing an allergic reaction among unexposed but sensitised subjects is notable and has to be considered. This aspect is even more frequent in the Mediterranean population among patients with vegetable allergies.

Concomitant sensitisation to pollen has always been observed in goji berry-sensitised individuals. Positive correlation was observed with typical Mediterranean sensitiser pollens including olive, *Artemisia*, *Cupressus* or *Platanus*. Additionally, latex allergy has also been associated with goji berries sensitisation and the authors even suggest that it may be a new player in latex-food syndrome (Gámez et al. 2013). Interestingly, concomitant co-sensitisation was observed in a cannabis-smoker who reported clinical symptoms after smoking cannabis (Larramendi et al. 2012).

In summary, goji berry sensitisation appears to be generally associated with other foods and pollen sensitisation and, although goji berries seem to be a potent allergenic source, it appears that sensitisation may be related to cross-reactivity with other frequently consumed sensitising food. Identifying the responsible proteins or allergens, when different food and pollens share specific allergens and, finally, whether cross-reactivity is mediated by these common allergens still must be elucidated.

12.10   Protein and Allergenic Composition of Goji Berries

Goji berries are extremely rich in nutrients. They are mainly composed of polysaccharides, which represent approximately 25–70 % of their total composition; carotenoids; large quantities of vitamins, especially vitamin C; proteins; fatty acids; free amino acids constituents; and minerals (Mikulic-Petkovsek et al. 2012). To date, the complete protein composition of goji berries has not been elucidated or published. Moreover, only partial studies have investigated the most relevant proteins, mainly associated with their healthy properties (Peng et al. 2012; Chen et al. 2014). It has been estimated that proteins represent approximately 12–14 % of their total composition. As for other fruits, their composition, especially with regard to the content and composition of vitamins and proteins, varies significantly depending on whether they are fresh or dried (Chang and So 2008).

From an allergenic perspective, proteins are the most relevant components, and they should be carefully extracted in order to maintain their characteristics. The allergen extracts are prepared with the dual objective of optimising the collection of responsible allergenic components and of studying the allergenic composition. Raw materials are usually homogenised and extracted under the appropriate conditions and using the appropriate buffers in order to maximise the extraction of allergenic proteins and to remove other components. By using aqueous extraction in a buffer solution containing phosphate-buffered saline/polyvinylpolypyrrolidone and freeze-drying the resultant solution, the protein content of goji berries yielded values of about 500 µg/g of freeze-dried material. Allergenic and non-allergenic proteins are contained in this extract.

The analysis of the protein composition, conducted by one- or two-dimensional (2D) electrophoretic assays, may be clearly identified. In the first case, different proteins are visualised in the 7–95 kDa range. The most prominent bands are detected at 7, 26, and 65 kDa. In 2D electrophoresis studies, more than 200 proteins were identified. The molecular weight of the proteins concurred with those observed in one-dimensional studies. The distribution of the proteins separated by the isoelectric point (pI) ranged in pH between 4 and 9. The 26 kDa band presented at least four isoforms. Some other bands were clearly visualised, especially those of 32 and 40 kDa. In general terms, all the proteins coincided with an acid pI (Fig. 12.1).

In this case, the protein profile does not match the allergenic profile of the extracts. For that reason, allergens are only described once they are immunologically

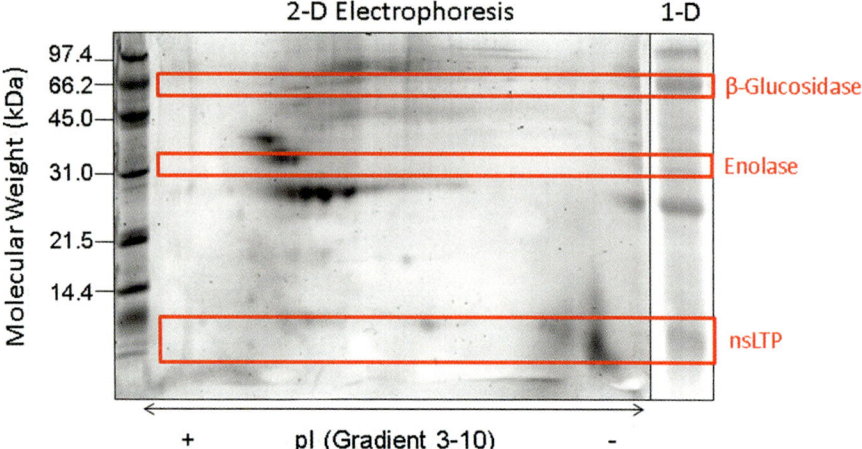

Fig. 12.1 Protein composition of goji berry allergenic extract. *2-D electrophoresis* shows different spots corresponding to proteins separated by molecular weight and isoelectric point. *1-D electrophoresis* shows proteins separated by molecular weight. The three allergens described and sequenced, with their corresponding molecular weight are framed in the figure. *nsLTP* non-specific lipid transfer proteins, *pI* isoelectric point

recognised by sensitised individuals who show specific IgE to the offending allergen extracts under study. In order to perform this evaluation, sera from sensitised individuals are required. It is also known that the allergenic profile can differ depending on the populations included in the study and each population may recognise different allergens depending on their geographical distribution.

Until now, no specific allergens from goji berries have been included in the allergen database (allergen.org). However, various published studies have identified different proteins which are clearly recognised by patients with severe symptoms following the ingestion of goji berries. The first study which investigated the allergenic composition of goji berries was published in 2013 (Carnés et al. 2013). Using patients from the Spanish Mediterranean region, the authors investigated the allergenic profile of an extract prepared from goji berries. The results obtained with a pool of sera showed the presence of eight different allergens. The study of individual patients demonstrated the existence of different allergens at 7, 25, 30, 37, 40, 46, 55 and 70 kDa, most of them being identified as major allergens (recognised by more than 50 % of the population). Monzón Ballarín et al. (2011) suggested the allergenic capacity of an nsLTP to be about 9 kDa and Larramendi et al. (2012) confirmed the importance of LTPs and identified a new allergen of about 50 kDa (Fig. 12.2).

In 2013, Carnés et al. characterised in detail the first goji berry allergen. This allergen was identified by liquid chromatography–mass spectrometry/mass spectrometry and was partially sequenced. The four fragments identified corresponded to the C-terminal of the protein. The authors concluded that the allergen corresponds

Fig. 12.2 Immunoblot hybridized with a pool of sera from goji berry-sensitised individuals. IgE-recognised bands are highlighted and signalled with an *arrow*. The allergenic profile of individuals demonstrate the recognition of several bands responsible for allergic sensitisation

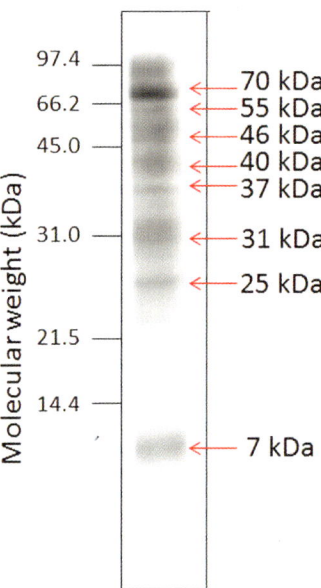

to a basic protein of 7 kDa and classified it in the group of non-specific LTPs. The alignment with other nsLTPs from other organisms shows 65 % homology with Pru p 3 (nsLTP from peaches), and 85 % with Sola l 3 (nsLTP from tomatoes) and nsLTP from *Nicotiana tabacum*. The allergen is registered in the international UniProtKB database (http://www.uniprot.org/) with the accession number B3A0N2, and the authors tentatively propose the name of Lyc bar 3 for this new allergen (Fig. 12.3).

In 2013, Gamez et al. published a study identifying two new allergens by matrix-assisted laser desorption/ionisation (MALDI)-time-of-flight (TOF) mass spectrometry (MS), consisting of a 60 kDa protein, identified as a β-glucosidase and a 35 kDa protein identified as an enolase. Both allergens were also recognised by a Spanish serum sample (Gámez et al. 2013).

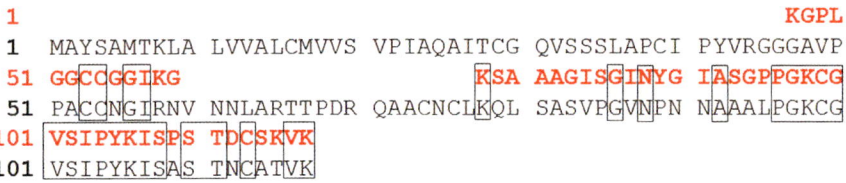

Fig. 12.3 Goji berry nsLTP aminoacid sequence of the four peptides identified (red). The sequence is aligned with the whole sequence of Pru p 3 (ns LTP from peach) (black). Common amino acids are framed. Partial sequence of nsLTP from goji berries shows a score of 65 % of homology with Pru p 3

12.11   Cross-Reactivity Between Goji Berries and Concomitant Sensitising Food

According to the allergen composition of goji berries and the allergic profile of patients investigated, various studies have suggested that there are several common allergens in goji berries and other fruits. These may be responsible for cross-reactivity between them.

Using in vitro experiments, Carnés et al. demonstrated cross-reactivity between goji berries and other foods and pollens, mediated by nsLTPs. In these studies, the incubation of serum samples from patients sensitised to goji berries with different foods containing high quantities of LTPs, including tomato peel, peach, a mixture of nuts (20% hazelnut, 20% almond, 20% peanut, 20% walnut and 20% coconut) or Artemisia pollen, inhibited the recognition of goji berry nsLTP (Fig. 12.4). These results were confirmed when serum samples were also inhibited with purified nsLTP from peaches (Pru p 3) and tomatoes (Sola l 3), confirming that nsLTP from goji berries, peaches or tomatoes share common IgE epitopes and explaining the high prevalence of co-sensitisation with other fruits (Fig. 12.5). The implication of nsLTPs in cross-reactivity was also previously pointed out by Larramendi et al. in vitro studies using serum samples from five patients.

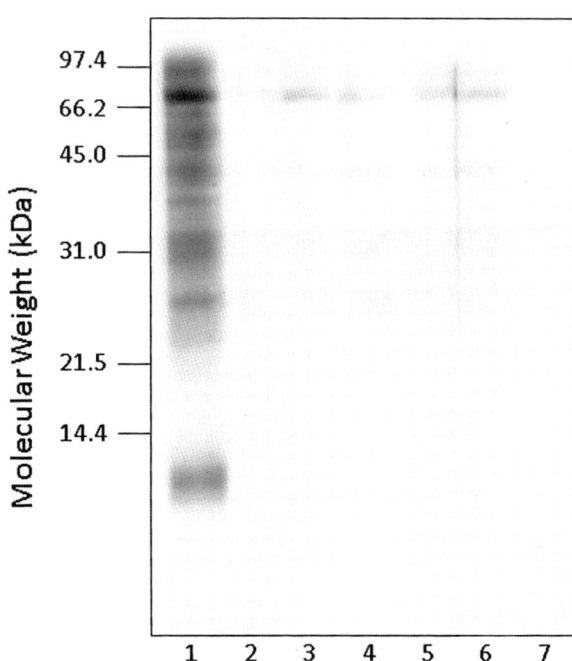

Fig. 12.4 Cross-reactivity between goji berries and different extract. Goji berry extract was inhibited with itself (lane *2*), peach extract (lane *3*), tomato extract (lane *4*), mixture of nut extracts (lane *5*) and *Artemisia vulgaris* pollen extract (lane *6*). Lanes *1* and *7* correspond to positive and negative control, respectively. The results show the capacity of different food extracts to inhibit the recognition of different bands in goji berry extracts confirming that concomitant sensitisation is a relevant factor in goji berry sensitised individuals

Fig. 12.5 Cross-reactivity between nsLTP of goji berries and peach. Lane *1* shows the immunoblot of goji berries hybridised with a positive pool of sera from goji berry-sensitised individuals. Lane *2* shows the same experiment but inhibiting previously the pool of sera with Pru p 3 (nsLTP from peach). The figure shows how Pru p 3 is capable to inhibit the recognition of nsLTP from goji berries demonstrating a similar pattern of IgE recognition and cross-reactivity between them. *nsLTP* non-specific lipid transfer proteins

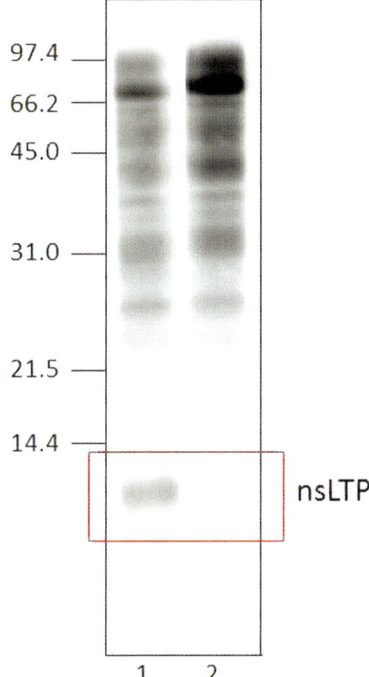

Moreover, in the previously explained investigations, authors have not only shown cross-reactivity with LTPs but also with other allergens present in the allergenic extracts. Unfortunately, neither of the authors investigated further into the identification of the other allergens involved, but both demonstrated that the incubation of goji berry-sensitised serum samples with different food and pollen extracts inhibited the recognition of goji berry allergens. Both studies concluded that there are other high-molecular-weight allergens that may also play an important role in cross-reactivity.

Additionally Guspi et al. (2011) suggested the importance of the 40 kDa protein, similar to an allergen present in potato. Gámez et al. (2013), suggested cross-reactivity between goji berries and latex in a Spanish patient. In spite of the fact that the authors demonstrated IgE binding of the serum sample to typical latex allergens, inhibition assays with goji berry extract must be performed in order to confirm cross-reactivity.

According to the data available in the literature, inhibition studies confirm the existence of cross-reactivity between goji berries and other fruits, nuts and vegetables. Based on these results, although other allergens are clearly involved in cross-reactivity mechanisms, to date only LTPs have been clearly identified and their role demonstrated. Further studies and the study of epitopes are needed to understand the mechanisms.

12.12 Non-IgE-Mediated Adverse Reactions

In general terms, goji berry toxicity has been considered low (Bensky et al. 2004). Recently, two cases of adverse reactions to goji berries have been described, both of them also in Spain. The first case is of a 60-year-old woman who developed a generalised erythematous and pruriginous maculopapular rash, asthenia, arthralgia, non-bloody diarrhoea, colic abdominal pain and jaundice after consuming goji berry tea three times a day over the previous 10 days (one handful of berries per cup). The study demonstrated hepatotoxicity, with marked elevation of transaminases, and withdrawal of the berries resulted in complete recovery, with liver function tests returning to normal values (Arroyo-Martínez et al. 2011). The presence of rash and systemic symptoms could be related to a delayed immune response. The second case is of a 53-year-old male with a pruriginous eruption located on sun-exposed areas of 2 weeks of duration while consuming goji berries for 5 months and infusions of cat's claw herb for 3 months, respectively. The minimal erythema dose for UVB (MED-UVB) was found to be decreased, and became normal when both products were withdrawn. Photo-provocation tests with goji berries and cat's claw were performed; the goji berries test proved positive, and systemic photosensitivity to goji berries was diagnosed (Gómez-Bernal et al. 2011).

12.13 Summary

In sum, the published studies to date (studies limited to patients from the Mediterranean area) confirm that individuals allergic to goji berries usually suffer from concomitant allergies, including allergies to fruits, nuts, other vegetables and even pollens, which are responsible for the primary sensitisation. Symptomatology following goji berry ingestion is mediated by cross-reactivity with concomitant food allergens. nsLTP, a family of panallergens with moderate cross-reactivity among different foods and pollens, are present in goji berries in high quantities as well as in concomitant food and pollen allergens. That means that allergy to goji berries, at least in the studied Mediterranean populations, can be explained by cross-reactivity between nsLTPs, which is the allergen responsible for the severe symptoms experienced by patients.

References

Arnold JJ, Williams PM. Anaphylaxis: recognition and management. Am Fam Physician. 2011;84:1111–8.
Arroyo-Martinez Q, Sáenz MJ, Argüelles Arias F, Acosta MS. *Lycium barbarum*: a new hepatotoxic "natural" agent? Dig Liver Dis. 2011;43:749.
Bacal LR. The impact of food allergies on quality of life. Pediatr Ann. 2013;42:141–5.

Beitia J, López-Matas MA, Alonso A, Vega A, Mateo B, Cárdenas R, Carnés J. Immunological efficacy after treatment with grass allergen-specific immunotherapy in patients diagnosed by component-resolved diagnosis. Int Arch Allergy Immunol. 2014;165:9–17.

Bensky D, Steven CS, Stöger E. Herbs that tonify the blood. In: Bensky D, Steven C, Stöger E, editors. Chinese herbal medicine: material medica. 3rd edn. Seattle: Eastland Press; 2004. pp. 743–66.

Bohle B. Impact of pollen-related food allergens on pollen allergy. Allergy. 2007;62:3–10.

Boyano-Martínez T, Pedrosa M, Belver T, Quirce S, García-Ara C. Peach allergy in Spanish children: tolerance to the pulp and molecular sensitization profile. Pediatr Allergy Immunol. 2013;24:168–72.

Cadavid Moreno S, Vicario García S, Bartolomé Zavala B, Morales Rubio C, Peláez Hernández A. Alergia alimentaria a bayas de Goji (*Lycium barbarum*). XXVII Congreso Nacional de la Sociedad Española de Alergología e Inmunología Clínica; 10–13 Nov. 2010; Madrid: J Investig Allergol Clin Immunol. p. 183.

Carnés J, Ferrer A, Huertas AJ, Andreu C, Larramendi CH, Fernández-Caldas E. The use of raw or boiled crustacean extracts for the diagnosis of seafood allergic individuals. Ann Allergy Asthma Immunol. 2007;98:349–54.

Carnés J, de Larramendi CH, Ferrer A, Huertas AJ, López-Matas MA, Pagán JA, Navarro LA, García-Abujeta JL, Vicario S, Peña M. Recently introduced foods as new allergenic sources: sensitisation to Goji berries (*Lycium barbarum*). Food Chem. 2013;137:130–5.

Chang RC, So KF. Use of anti-aging herbal medicin, Lycium barbarum, against aging-associated diseases. What do we know so far? Cell Mol Neurobiol. 2008;28:643–52.

Chen KC, Chen KB, Chen HY, Chen CY. In silico investigation of potential pyruvate kinase M2 regulators from traditional Chinese medicine against cancers. Biomed Res Int. 2014:189495.

Gámez C, Marchán E, Miguel L, Sanz V, del Pozo V. Goji berry: a potential new player in latex-food syndrome. Ann Allergy Asthma Immunol. 2013;110:206–7.

Gómez-Bernal S, Rodríguez-Pazos L, Martínez FJ, Ginarte M, Rodríguez-Granados MT, Toribio J. Systemic photosensitivity due to Goji berries. Photodermatol Photoimmunol Photomed. 2011;27:245–7.

Guspi R PF, Baltasar M. A new allergenic fruit: urticarial after eating goji berries. XXX Congress of the European Academy of Allergy and Clinical Immunology; 11–15 June 2011; Istanbul: Allergy. p. 325.

Himmel ME, Yao Y, Orban PC, Steiner TS, Levings MK. Regulatory T-cell therapy for inflammatory bowel disease: more questions than answers. Immunology. 2012;136:115–22.

Knibb RC, Semper H. Impact of suspected food allergy on emotional distress and family life of parents prior to allergy diagnosis. Pediatr Allergy Immunol. 2013;24:798–803.

Larramendi CH, Ferrer A, Huertas AJ, García-Abujeta JL, Andreu C, Tella R, Cerdà MT, Bartra J, Lavín JR, Pagán JA, López-Matas MA, Fernández-Caldas E, Carnés J. Sensitization to tomato peel and pulp extracts in the Mediterranean coast of Spain: prevalence and co-sensitization with aeroallergens. Clin Exp Allergy. 2008;38:169–77.

Larramendi CH, García-Abujeta JL, Vicario S, García-Endrino A, López-Matas MA, García-Sedeño MD, Carnés J. Goji berries (*Lycium barbarum*): risk of allergic reactions in individuals with food allergy. J Investig Allergol Clin Immunol. 2012;22:345–50.

Mikulic-Petkovsek M, Schmitzer V, Slatnar A, Stampar F, Veberic R. Composition of sugars, organic acids, and total phenolics in 25 wild or cultivated berry species. J Food Sci. 2012;77:C1064–70.

Monzón Ballarín S, López-Matas MA, Sáenz Abad D, Pérez-Cinto N, Carnés J. Anaphylaxis associated with the ingestion of Goji berries (*Lycium barbarum*). J Investig Allergol Clin Immunol. 2011;21:567–70.

Morales M, López-Matas MA, Moya R, Carnés J. Cross-reactivity among non-specific lipid-transfer proteins from food and pollen allergenic sources. Food Chem. 2014;165:397–402.

Muraro A, Werfel T, Hoffmann-Sommergruber K, Roberts G, Beyer K, Bindslev-Jensen C, Cardona V, Dubois A, duToit G, Eigenmann P, Fernandez Rivas M, Halken S, Hickstein L, Høst A, Knol E, Lack G, Marchisotto MJ, Niggemann B, Nwaru BI, Papadopoulos NG, Poulsen LK,

Santos AF, Skypala I, Schoepfer A, Van Ree R, Venter C, Worm M, Vlieg-Boerstra B, Panesar S, de Silva D, Soares-Weiser K, Sheikh A, Ballmer-Weber BK, Nilsson C, de Jong NW, Akdis CA, EAACI Food Allergy and Anaphylaxis Guidelines Group. EAACI food allergy and ana- phylaxis guidelines: diagnosis and management of food allergy. Allergy. 2014a;69:1008–25.

Muraro A, Dubois AE, DunnGalvin A, Hourihane JO, de Jong NW, Meyer R, Panesar SS, Roberts G, Salvilla S, Sheikh A, Worth A, Flokstra-de Blok BM. EAACI food allergy and anaphylaxis guidelines. Food allergy health-related quality of life measures. Allergy. 2014b;69:845–53.

Nowak-Węgrzyn A, Albin S. Oral immunotherapy for food allergy: mechanisms and role in man- agement. Clin Exp Allergy. 2014 DOI: 10.1111/cea.12382.

Nwaru BI, Hickstein L, Panesar SS, Muraro A, Werfel T, Cardona V, Dubois AE, Halken S, Hoff- mann-Sommergruber K, Poulsen LK, Roberts G, Van Ree R, Vlieg-Boerstra BJ, Sheikh A, EAACI Food Allergy and Anaphylaxis Guidelines Group. The epidemiology of food allergy in Europe: a systematic review and meta-analysis. Allergy. 2014;69:62–75.

Pajno GB, Cox L, Caminiti L, Ramistella V, Crisafulli G. Oral immunotherapy for treatment of immunoglobulin E-mediated food allergy: the transition to clinical practice. Pediatr Allergy Immunol. 2014;27:42–50.

Peng Q, Song J, Lv X, Wang Z, Huang L, Du Y. Structural characterization of an arabinogalactan- protein from the fruits of *Lycium ruthericum*. J Agric Food Chem. 2012;60:9424–9.

Radauer C, Willerroider M, Fuchs H, Hoffmann-Sommergruber K, Thalhamer J, Ferreira F, Scheiner O, Breiteneder H. Cross-reactive and species-specific immunoglobulin E epitopes of plant profilins: an experimental and structure-based analysis. Clin Exp Allergy. 2006;36:920–9.

Rescigno M. Dendritic cells in oral tolerance in the gut. Cell Microbiol. 2011;13:1312–8.

Schulten V, Lauer I, Scheurer S, Thalhammer T, Bohle B. A food matrix reduces digestion and absorption of food allergens in vivo. Mol Nutr Food Res. 2011;55:1484–91.

Song TT, Worm M, Lieberman P. Anaphylaxis treatment: current barriers to adrenaline auto-injec- tor use. Allergy. 2014;69:983–91.

Tripodi S, Frediani T, Lucarelli S, Macrì F, Pingitore G, Di Rienzo Businco A, Dondi A, Pansa P, Ragusa G, Asero R, Faggian D, Plebani M, Matricardi PM. Molecular profiles of IgE to *Ph- leum pratense* in children with grass pollen allergy: implication for specific immunotherapy. J Allergy Clin Immunol. 2012;129:834–9.

Printed by Printforce, the Netherlands